MATH REVIEW TOOLKIT

GARY L. LONG
Virginia Tech

SHARON D. LONG

GENERAL CHEMISTRY

FOURTH EDITION

ILL • PETRUCCI • McCREARY • PERRY

Upper Saddle River, NJ 07458

Project Manager: Kristen Kaiser
Senior Editor: Kent Porter-Hamann
Editor-in-Chief, Science: John Challice
Vice President of Production & Manufacturing: David W. Riccardi
Executive Managing Editor: Kathleen Schiaparelli
Assistant Managing Editor: Becca Richter
Production Editor: Dana Dunn
Supplement Cover Manager: Paul Gourhan
Supplement Cover Designer: Joanne Alexandris
Manufacturing Buyer: Ilene Kahn

PEARSON Prentice Hall

Pearson Prentice Hall
Pearson Education, Inc.
Upper Saddle River, NJ 07458

Printed in the United States of America

10 9 8 7 6 5 4 3 2 1

ISBN 0-13-150824-5

Pearson Education Ltd., *London*
Pearson Education Australia Pty. Ltd., *Sydney*
Pearson Education Singapore, Pte. Ltd.
Pearson Education North Asia Ltd., *Hong Kong*
Pearson Education Canada, Inc., *Toronto*
Pearson Educación de Mexico, S.A. de C.V.
Pearson Education—Japan, *Tokyo*
Pearson Education Malaysia, Pte. Ltd.

Contents

Preface ... vii

1 Harnessing the Power of Your Calculator ... 1
- 1.1 Ten-Dollar Calculators That Work ... 1
- 1.2 The First Five Buttons–The Basic Operations ... 2
- 1.3 Rounding Off and Significant Figures ... 4
- 1.4 What Are the Rules for Significant Figures? ... 5
- 1.5 Exponential Notation to the Rescue ... 7
- 1.6 Rules for Rounding Off ... 8
- 1.7 Significant Figures in Multiplication and Division ... 9
- 1.8 [EXP]: Exponential Notation on the Calculator ... 9
- 1.9 Multiplication and Division with Exponential Notation ... 11
- 1.10 Switching Between Exponential Notation and Decimal ... 12
- 1.11 Does the Answer Make Sense? ... 18
- 1.12 Practice Problems ... 20
- 1.13 Answers to Practice Problems ... 21

2 Skills for Chapter 1: Chemistry: Matter and Measurement ... 22
- 2.1 Significant Figures ... 22
- 2.2 Exponential Notation ... 23
- 2.3 Mean and Standard Deviation ... 24

3 Skills for Chapter 3: Stoichiometry: Chemical Calculations ... 26
- 3.1 Yields ... 26
- 3.2 Molarity ... 27
- 3.3 Dilution ... 29
- 3.4 Titration ... 30

4 Skills for Chapter 5: Gases ... 32
- 4.1 Gas Laws ... 32
- 4.2 Ideal Gas Law ... 33
- 4.3 Dalton's Law ... 35
- 4.4 Graham's Law ... 36

5 Skills for Chapter 6: Thermochemistry ... 38
- 5.1 Calorimetry ... 38
- 5.2 Standard Enthalpy of Reaction ... 39

6 Skills for Chapter 7: Atomic Structure 41
6.1 Radiant Energy 41
6.2 Bohr Energy-Level Equation 42
6.3 Wavelike Properties of Matter 43

7 Skills for Chapter 9: Chemical Bonds 45
7.1 Bond Dissociation Energies 45

8 Skills for Chapter 12: Solutions 46
8.1 Concentration Units 46
8.2 Colligative Properties 46

9 Skills for Chapter 13: Chemical Kinetics 48
9.1 Reaction Rates 48
9.2 Concentration and Time 50
9.3 Reaction Rates and Temperatures 54

10 Skills for Chapter 14: Chemical Equilibrium 55
10.1 Calculations with Equilibrium Expressions 55
10.2 Calculations of Equilibrium Concentrations 57

11 Skills for Chapter 15: Acid-Base Equilibria 63
11.1 pH Calculations 63
11.2 Equilibrium Calculations 65
11.3 Common Ion Effect 69
11.4 Buffers 72

12 Skills for Chapter 16: Equilibria Involving Slightly Soluble Salts and Complex Ions 78
12.1 Solubility 78
12.2 Effect of Common Ions and pH 81
12.3 Complex Ions 81

13 Skills for Chapter 17: Thermodynamics 85
13.1 Enthalpy and Entropy Changes 85
13.2 Gibbs Free Energy 85
13.3 Free Energy and Equilibrium 86

14 Skills for Chapter 18: Electrochemistry *88*
14.1 emf, *K*, and the Nernst Equation *88*
14.2 Electrolysis *92*

15 Skills for Chapter 19: Nuclear Chemistry *93*

16 Self Test for Math Skills *95*

17 Chemistry and Writing *99*
17.1 The Scientific Notebook *99*
17.2 The Scientific Report *101*
17.3 Technical Writing *103*
17.4 A Notebook Example *106*

18 Chemistry and Career Planning *110*
18.1 Materials Science *110*
18.2 Polymer Science *111*
18.3 Environmental Science and Technology *111*
18.4 Biochemistry and Biotechnology *112*
18.5 Medicinal Chemistry and Clinical Chemistry *112*
18.6 Forensic Science *113*
18.7 Radiochemistry *114*
18.8 Patent Attorneys and Patent Agents *114*
18.9 Chemical Education *115*
18.10 Chemical Information Careers *115*
18.11 Other Chemistry Careers *116*

19 Study Aids for Chemistry *117*
19.1 Good Study Habits *118*
19.2 Reading Your Textbook *119*
19.3 Learning to Relax *121*
19.4 Conquering Tests *123*

20 Math Skills Test Answers *127*

Acknowledgement

The authors (GLL and SDL) would like to acknowledge the contributions of Doris L. Lewis on Sections 17 and 18 of this Toolkit on Chemistry and Writing, and Careers in Chemistry. Also acknowledged are the contributions of Timothy Smith and Diane Vukovich on Sections 1 and 18 concerning Calculators and Study Aids for Chemistry.

PREFACE To The Student

A successful study of general chemistry requires you to use your skills of memorization, basic logic and mathematics. From my years in the classroom I have found that students who are unsure about their math skills generally do not do well on their exams. To put this statement into context, imagine that you are holding a chemistry exam containing 20 multiple choice questions, with 15 of them involving the use of mathematical equations. If you have trouble with the first few questions involving math and have only an hour to complete the exam, any weakness in math skills may limit your performance on the exam, regardless of how many facts and figures you have memorized.

The math skills that a student needs to successfully complete general chemistry are basically what you have learned in high school; that is, algebra and trigonometry. Calculus is not required. It is the purpose of this booklet to guide you chapter by chapter in the mathematics that are used in the fourth edition text of GENERAL CHEMISTRY. The math skills for each chapter are outlined and examples are worked in this booklet.

The case of a life sciences student planted the idea for this booklet. This young man was going around for the seventh time in general chemistry. He could never manage to pass the first test so he would drop out and wait until the next semester to try again. Fortunately, he came to me at the beginning of the course. He was desperate; he could not graduate until he passed chemistry. After speaking with him, I found his limitation was not inadequate math training, but a basic fear of math problems. He would "freeze up" on the math problems on the exam. Over the course of the semester I met with the student and explained how to perform these calculations using the methods that are described in this booklet. The student successfully completed the course with a B average.

Although we cannot absolutely guarantee your success in chemistry with the use of this booklet, mastery of the material presented here will enhance your ability to work problems in general chemistry. Take a few moments and complete the SELF TEST FOR MATH SKILLS at the back of this booklet. It will help you assess your skills. Use this guide to help you reinforce your math skills.

Chemistry is an exciting science that touches every aspect of our lives. It is our hope that this booklet may demystify the mathematics used in general chemistry and help you discover this excitement.

Gary L. Long
long@vt.edu

I will never forget my own struggle with chemistry at Wake Forest University. As a freshman, I had not yet developed productive study habits. I faced every new chemistry chapter with a certain dread. There is a tendency to procrastinate taking hold of something we fear or find distasteful, or so it was in my case. However, if you cope with chemistry in this manner, as I did, the stress of catching up and preparing for a test will cost you your well being until the test is over. The two grades of C that I made in freshman chemistry taught me this painful lesson. The point being, even if you dislike chemistry and are just taking it only to satisfy the requirements for your major, make your effort consistent.

Many college students have been forced to change their career dreams because they could not make it through chemistry. This unpleasantness does not give this science a good name nor does it give chemistry professors an easy grace at social functions when asked their occupation. Some of you may need to change your career option to a different discipline for which you are gifted. A very kind English professor pointed this out to me at a time when organic chemistry was causing me great distress. Time is too valuable to spend it in the wrong place. However, for those of you heading in the right direction, we do not want chemistry to be a stumbling block for your dreams.

We hope that your experience with chemistry will be a victorious one and that in some small way this math book will make a difference. Write us and tell us your stories as well as your triumphs.

Sharon D. Long
sdlong@vt.edu

Section 1

Harnessing the Power of Your Calculator

The best friend you have in a chemistry class is your calculator. If used correctly, it will save you precious seconds during a test, make doing homework problems outside class much more efficient, and allow you to spend more time on the concepts rather than laboring over the calculations.

In your general chemistry course, you will encounter problems using arithmetic operations (addition, subtraction, multiplication, and division), trigonometric operations (cos, sin, $\cos^{-1}$, and $\sin^{-1}$), and logarithmic operations (log, ln, 10^x, e^x, powers and roots). You will also deal with exponential notation in many of these problems. While your textbook has an excellent appendix on mathematics for general chemistry, we present in this tool kit methods for using your calculator to solve these calculations.

Take the time to acquaint yourself with the capabilities of your calculator. Please try the practice problems in Section 1.12 of this tool kit. You will find your calculator to be a powerful tool that you can harness for your study of general chemistry.

1.1 Ten-Dollar Calculators That Work

Many students have difficulty in the first few weeks of the semester because they don't have the type of calculator they need in order to quickly and efficiently perform the tasks the instructor and textbook will require of them. You may sit down in your first class with a four-function calculator that just does the basic operations of adding, subtracting, multiplying, and dividing (the type the bank gives you when you open a checking account). This type of calculator is not powerful enough to do many of your chemistry problems. On the other hand, you don't want to purchase an engineering calculator that has so many functions it is intimidating to look at and has a manual as thick as *The Hobbit!*

You need a scientific calculator that will help you quickly and easily dispatch the problems set before you in your class, especially on tests. The calculator should have the following capabilities:

- The four basic arithmetic functions.
- A button to enter exponential notation (EE or EXP on most calculators).
- Buttons to easily enter logarithms.
- Buttons to convert between exponential notation and decimal form.
- Buttons to do statistical calculations. (You won't need them in this class, but many of you will probably take statistics later in your program.)
- At least 10 places on the display so you can display fairly large numbers in decimal form. (Called 10 + 2 on some models.)

Below is a list of calculators that are inexpensive (around $10) and do the job you need a calculator to do.

- Casio fx-250HC
- Texas Instruments TI-30X family

1.2 The First Five Buttons—The Basic Operations

Nothing is easier than multiplying or dividing on a calculator. If you want to multiply 2.71 × 1.91, you simply punch

2.71 [×] 1.91 [=]　　　　The display reads 5.1761

When you divide two numbers, the process is the same. Let's divide 3.75 by 1.2. This problem can be presented to you as 3.75 [÷] 1.2 or 3.75/1.2 or $\frac{3.75}{1.2}$. No matter how the division is written, it is entered into the calculator as

3.75 [÷] 1.2 [=] The display reads 3.125

Most of the time in chemistry you will do several operations in one problem. An example of this type of problem is

$$2.875 \times \frac{1}{3.125} =$$

Many students enter the operations like this:

2.875 [×] 1 [÷] 3.125 [=] The display reads 0.92

It is much simpler and less prone to error if the multiplication step is left out when all you are doing is multiplying by one. You simply enter

2:875 [÷] 3.125 [=] The display reads 0.92

Later, when you do conversions, you will do many multiplications and divisions in one problem. Let's try one.

$$\frac{2.375 \times 4.15}{0.03125 \times 1.577} =$$

Many students do the problem like this:

First, they multiply the numbers in the numerator

2.375 [×] 4.15 [=] The display reads 9.85625

Then they write this answer down, push [C] to clear the display, and then multiply the numbers in the denominator:

0.03125 [×] 1.577 [=]
The display reads 0.04928125

Then they write this number down, clear the display, reenter 9.85625, and finally divide by 0.04928125 to get the answer 200.

In other words, they multiply the numbers in the numerator first, erase the answer, then multiply the numbers in the denominator, reenter the first answer, and then do the division.

There is an easier way to solve this equation. Do the multiplications and divisions in order from left to right. Press [×] before numbers that are in a numerator (after the first number) and press [÷] before any number in a denominator.

2.375 [×] 4.15 [÷] 0.03125 [÷] 1.577 [=]
The display reads 200

You can change the order, as long as you press [×] before a number in the numerator and press [÷] before a number in the denominator.

2.375 [÷] 0.03125 [÷] 1.577 [×] 4.15 [=]
The display reads 200

Let's try another.

$$1.44 \times \frac{1}{16} \times \frac{13.3}{0.20} =$$

We punch

1.44 [÷] 1.6 [×] 13.3 [÷] 0.20 [=]
The display reads 5.985

Doing problems this way can save you much time on tests.

1.3 Rounding Off and Significant Figures

If you divide 2 by 7 on your calculator, the display reads 0.285714286. Ten decimal places can be quite cumbersome. Fortunately, you do not have to write all those numbers when you answer a problem; you can round off your answer. Answers beyond a certain number of places are not just cumbersome, they are misleading. So we round our answers off to a reasonable number of decimal places.

In order to round off, we have to understand significant figures (sometimes referred to as sig figs). In chemistry (and in all science), we do not work with pure numbers as in algebra class. We work with measurements. If you take your body temperature with a fever thermometer, it may look like the figure below. The mercury stops between 98 and 99 degrees. It looks like it is between 98.2 and 98.4 degrees (each small line on the thermometer is 0.2 of one degree), so we might guess 98.3. In all measurements, the last digit is a good guess and your guess might be a little different from mine. This temperature (98.3 degrees) has three significant figures. If we round it off to two places, it will have two significant figures (98 degrees).

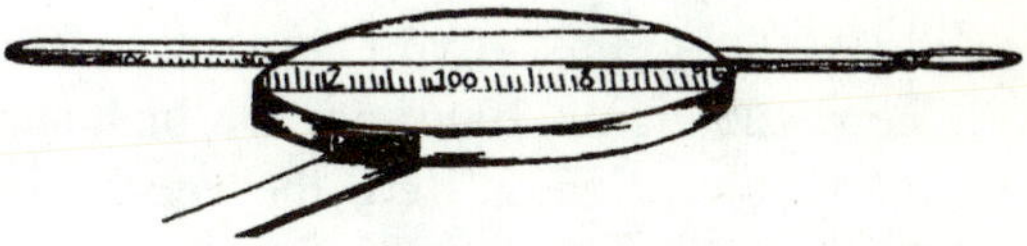

A thermometer showing a reading of 98.3 degrees.

Some laboratory balances read to the hundredths place. The digital readout of an object's mass might look like this: 14.49. This reading has four significant figures. Other (more expensive) balances read to the ten-thousandths place. The digital display on this balance might read 14.4927 for this object. This reading has six significant figures.

1.4 What Are the Rules for Significant Figures?

How can you tell if digits are significant? Most of the confusion involves the number zero. Your textbook gives you rules similar to these.

All non–zero digits are significant.

This means that any number that is not zero is significant. 7.121486 has seven significant figures; 4215 has four significant figures.

All zeros between non–zero digits are significant.

7002 has four significant figures; 7.1002 has five significant figures; 702 has three significant figures; 80.003 has five significant figures.

All zeros to the left of the first non–zero digit are not significant.

Here is where your normal everyday definition of "significant" may clash with the chemistry definition. The number 0.0025 has only two significant figures, as the zeros to the left of the two are not significant. That does not mean they don't have to be there. If you took them away, you would get .25, a very different number! A common dosage of Vasotec™, a blood pressure-lowering drug, is 0.0025 grams. If you ignored the zeros, the dosage would become .25 grams, 100 times too high and probably fatal to the patient. Those zeros have the very important job of being placeholders; however, they are not significant in the sense of expressing the accuracy of the number.

If we convert 0.0025 kilograms to grams, it becomes 2.5 grams. Because we changed the units, we no longer needed the left-hand zeros as placeholders.

When a decimal point is present, zeros to the right of the last non-zero digit are significant.

Thus, 0.2300 has four significant figures, 2.4900 has five significant figures, and 2.0 has two significant figures.

If there is no decimal point written, you cannot tell if the zeros to the right are significant. This rule can confuse us if we are not careful.

For instance, we cannot say how many significant figures whole numbers like 8000 and 400 have. They have one significant figure; beyond that we cannot say.

Now let's look at some examples that involve all the rules.

- 7.0010 has five significant figures. (All the zeros are significant because two are between non-zero digits and the last is in a number that has the decimal point written.)

- 0.0005 has one significant figure. (The zeros to the left are not significant.)
- 0.000500 has three significant figures. (The zeros to the left of the 5 are placeholders but are not significant. The zeros to the right are significant because a decimal point is shown.)
- 1.000500 has seven significant figures. (All the zeros are significant because the three zeros are between non-zero digits and the last two are to the right in a number that has the decimal point written.)

- 7000 has one significant figure. (No decimal point is shown, so you cannot tell whether the zeros are significant.)

- 7000. has four significant figures. (This time the decimal point is shown.)

See Appendix A.4 of your text for a discussion on significant figures in logarithms.

1.5 Exponential Notation to the Rescue

Let's look at the number 800. Suppose it represents 800 grams. If 800 grams is just a good guess at the weight of an object, then the number has only one significant figure. However, if you weighed the object and it weighed exactly 800 grams, the number would have three significant figures. If you want to show that the number has three significant figures, you would have to write it with a decimal point included, that is, 800. grams. But it is easy for a decimal point at the end of a whole number to "get lost." However, we can easily show the number of significant figures in numbers such as this by converting them to exponential notation. If you convert 800 to exponential notation, it becomes 8×10^2.

If you want to show that it has three significant figures, you can write it as 8.00×10^2.

When you write a number in exponential notation, the number of significant digits is expressed by the coefficient.

- 7000 grams becomes 7×10^3 with one significant figure
- 7000 grams becomes 7.0×10^3 with two significant figures
- 7000 grams becomes 7.00×10^3 with three significant figures
- 7000 grams becomes 7.000×10^3 with four significant figures

Remember that the exponent plays no part in determining the number of significant figures. It merely tell us where to put the decimal point.

Appendix A.1 of your text further describes exponential notation and its use in arithmetic and multiplicative operations.

1.6 Rules for Rounding Off

You may remember these rules from a math class in high school. To round off a number to a certain place, you look at the number in the next place to the right of the number. If that number to the right is less than 5, that is, if it is 0, 1, 2, 3, or 4, you leave the number alone. This is called rounding down. If the number to its right is 5 or greater, that is, if it is 5, 6, 7, 8, or 9, you increase the value of the number by one. This is called rounding up.

Let's round some numbers.

- 7.134 rounded to the hundredths place becomes 7.13 since the number in the thousandths place is 4, a number less than 5.

- 0.21029 rounded to the ten-thousandths place becomes 0.2103 since the number to the right of that place is 9, a number that is 5 or greater.

1.7 Significant Figures in Multiplication and Division

When you multiply or divide measurements, the number with the least number of significant figures limits your answer. Suppose three students measure a box. Jane measures the length to be 3.12 cm, Carl measures the width to be 3.1 cm, and Donna measures the height to be 7.62 cm. To get the volume, we multiply the three sides together.

3.12 cm [×] 3.1 cm [×] 7.62 cm [=]
The display reads 73. 70064, with the answer in cm^3

However, we must round the calculated answer to 74 cm^3 because Carl made his measurement to only two significant figures.

When we calculate density, we divide mass by volume. That is,

$$\text{Density} = \frac{\text{Mass}}{\text{Volume}}$$

Let's say we measure a sample's mass to be 7.4219 grams and its volume to be 1.23 mL. We divide 7.4219 by 1. 23.

$$\text{Density} = \frac{7.4219 \text{ g}}{1.23 \text{ mL}}$$

$$\text{Density} = 6.034065041 \ {}^{\text{g}}\!/_{\text{mL}}$$

Because 1.23 mL has only three significant figures, we must round the answer to 6.03 g/mL.

Significant figures are important, but don't get too bogged down with them. They will come naturally to you as you do more calculations.

1.8 [EXP]: Exponential Notation on the Calculator

Calculators make exponential notation easy. You just need to have the right calculator and know which buttons to push.

Note: Scientific calculators have a button that is marked EE (on most Texas Instruments calculators) or EXP (on most other brands). This is the button that lets you enter a number in exponential notation.

Let's take the number 7.45×10^8. To enter this number, we enter the coefficient 7.45, then press [EXP]. Two zeros appear at the right (smaller in size). These zeros are holding the places for the two digits of the exponent. (You will never need more than two.) We then enter the exponent 8. The display should look like this: 7. 45 ^{08}

When you see this display, you read it as 7.45×10^8. Press [C] to clear the display, and let's enter it again.

7.45 [EXP] 8 The display reads 7. 45 ^{08}

Note: A common mistake students make is to hit [X] before [EXP]. Don't do this! The "times 10 to the" is all included in [EXP].

So you *do not* enter

7.45 [×] [EXP] 8 or 7.4501 [×] 10 [EXP] 8

Simply press 7.45 [EXP] 8

Another example is to enter 9.12×10^{12}

Press 9.12 [EXP] 12 The display reads 9. 12 ^{12}

For an entry with a negative exponent, try 7.216×10^{-9}

Enter 7.216 [EXP] 9 [±] or 7.216 [EXP] [±] 9
The display reads 7. 216 $^{-09}$

The [±] key toggles a number between negative and positive. If you press [±] after [EE] or [EXP], it toggles the exponent between negative and positive. Press [±] a couple of times and see how the sign of the exponent changes back and forth.

Note: The [±] key is not the same as [–]. The [±] toggles between negative and positive. [–] is used only for subtraction.

Let's enter a few more numbers in exponential notation. Enter the mass of an electron, which is 1.6606×10^{-24} kilograms.

Press 1.6606 [EXP] 24 [±]
The display reads 1.6606 $^{-24}$

The second example is to enter 10^5. On your calculator:

Press 1 [EXP] 5 The display reads 1.0 ^{05}

Note: When entering an exponential number where no coefficient is shown, you must supply the missing coefficient of 1. We can write 10^5 as 1×10^5.

Finally, let's enter 10^{-7}

Press 1 [EXP] 7 [±] The display reads 1.0 $^{-07}$

1.9 Multiplication and Division with Exponential Notation

If you multiply two numbers in exponential notation, just press [×] between the numbers. Let's multiply $(2.12 \times 10^4)\ (7.89 \times 10^{-9}) =$

Enter 2.12 [EXP] 4 [×] 7.89 [EXP] 9 [±] [=]
The display reads 1.67268 $^{-04}$

Since each number has three significant figures, we round the answer to 1.67×10^{-4}. Note that the only time you enter [×] is between the two numbers.

Let's divide two numbers: $\dfrac{8.81 \times 10^4}{7.472 \times 10^8}$

Enter 8.81 [EXP] 4 [÷] 7.472 [EXP] 8 [=]
The display reads 1.1790685 $^{-04}$

Since the least number of significant figures is three, we round our answer to 1.18×10^{-4}.

Let's try another. $\frac{10^{-14}}{10^{-9}}$ (Remember to supply the missing coefficients of 1.)

Press 1 [EXP] 14 [±] [÷] 1 [EXP] 9 [±] [=]
The display reads 1.0000000 $^{-05}$, written 1×10^{-5} or 10^{-5}.

Now let's look at one last example that uses Avogadro's number (6.02×10^{23}), a number that you will see later in the class.

$$2.50 \times 10^{15}\left(\frac{12.0}{6.02 \times 10^{23}}\right)$$

Press 2.5 [EXP] 15 [×] 12 [÷] 6.02 [EXP] 23 [=]
The display reads 4.9833887 $^{-08}$.

Since each of the three numbers has three significant figures, we round the answer to 4.98×10^{-8}.

1.10 Switching Between Exponential Notation and Decimal

Sometimes you may want to convert an answer back and forth between exponential notation and decimal form. This will be most important to you on exams, especially multiple-choice exams where the answer choices could be written in either form.

The answers to problems in the back of your textbook may also be given in either exponential notation or decimal form. Often students are confused when they look up a hard-fought answer in the answer key and discover their answer looks wrong. Their answer may not be wrong. The textbook may have given the answer in exponential notation, while the student calculated it in decimal form.

Texas Instruments Calculators

On Texas Instrument calculators, we use [FLO] and [SCI] along with [2nd] to convert between exponential notation and decimal form. On the TI-30X calculator, [FLO] and [SCI] are second functions written in blue above [4] and [5]. We use [2nd] to access these operations.

Pushing [2nd] [FLO] puts the calculator into normal decimal mode. (FLO comes from "floating point," computer programmers' jargon for a decimal number. In this mode, any number we enter will change to decimal form as soon as we hit another key (as long as the number will fit the display).

Pushing [2nd] [SCI] puts the calculator into exponential notation mode. Any number we enter will change to exponential notation as soon as we hit another key.

First let's change some numbers in exponential notation to decimal form. Suppose you got 7.7×10^{-4} as an answer and want convert it to decimal form.

Enter 7.7 [EE] 4 [±] The display reads 7. 7 $^{-04}$

Press [2nd] [FLO] The display reads 0. 00077

Now let's convert 8.449×10^{7} to decimal form.

Enter 8.449 [EE] 7 The display reads 8. 449 ^{07}

Press [2nd] [FLO] The display reads 84490000

Now let's try to convert 6.419×10^{-17} to decimal form.

Enter 6.419 [EE] 17 [±] The display reads 6. 419 $^{-17}$

Press [2nd] [FLO] The display still reads 6. 419 $^{-17}$

Nothing happens, because a number in exponential notation with an exponent of –17 has too many decimal places to fit into the 10 spaces on the calculator. A number in exponential notation must have an exponent of 9 or less, either plus or minus, to fit into the ten decimal places on a calculator. That is, a number must be between nine billion and nine-billionth.

Now let's take some decimal numbers and convert to exponential notation. Suppose you got 0.000788 for an answer and want to change it to exponential notation.

Enter 0.000788	The display reads 0. 000788
Press [2nd] [SCI]	The display reads 7. 88 $^{-04}$

Convert 6,992,000 to exponential notation.

Enter 6,992,000	The display reads 6992000
Press [2nd] [SCI]	The display reads 6. 992 ^{06}

You can press [2nd] [SCI] to convert back to normal decimal form.

Note: If you press [2nd] [SCI] to convert an answer to exponential notation, the calculator remains in that mode until you press [2nd] [FLO] to put the calculator into normal decimal mode. If you enter a decimal number, it will change to exponential notation as soon as you press another function key such as [×], [÷] or [=].

Likewise, if you are in decimal mode when you enter a number in exponential notation, it will change to decimal form as soon as you hit another key (as long as the decimal form will fit in the display). In either of these cases the number is not affected, only the form of the number.

Let's look at a few examples. Press [2nd] [SCI] to put the calculator in exponential notation mode

Enter 0.0059 The display reads 0.0059

Press [×] or [÷]

The display changes to 5.9000000^{-03}

Press [C] to clear the display. Now press [2nd] [FLO] to change your calculator to normal decimal mode.

Now enter 0.0059 The display reads 0.0059

Now press [×] or [÷] The display still reads 0.0059

Casio, Sharp, and Most of the Rest

Most brands of calculators have a [MODE] button, usually in the upper left hand corner next to [INV]. They also have a table printed under the readout display that looks something like this.

MODE	•	0	4	5	6	7	8	9
	SD	COMP	DEG	RAD	GRA	FIX	SCI	NORM

or this

MODE	0:COMP	4:DEG	5:RAD	6:GRA
	• SD	7:FIX	8:SCI	9:NORM

The two numbers you need in this table to convert between exponential notation and decimals are 8:SCI (exponential notation) and 9:NORM (for normal decimal form).

Pressing [MODE] [9] puts the calculator into normal decimal mode. In this mode, a number you enter will change to decimal form as soon as you hit another key (as long as the number will fit in the display).

Pressing [MODE] [8] [8] puts the calculator into exponential notation mode. Any number you enter will change to exponential notation as soon as you hit another key. The second number you enter after pressing [MODE] [8] tells the calculator how many places of the coefficient the calculator will display. For instance, if you press [MODE] [8] [4], the answer will be rounded to 4 digits. It's better to let the calculator display 8 digits; then you can choose the proper number of significant figures for an answer.

Let's change some numbers from exponential notation to normal decimal form. Suppose you got 7.7×10^{-4} as an answer and want to change it to decimal form.

Enter 7.7 [EXP] 4 [±] The display reads 7.7 $^{-04}$

Press [MODE] 9 The display reads 0.00077

Convert 2.568×10^{8} to decimal form.

Enter 2.568 [EXP] 8 The display reads 2.568 ^{08}

Press [MODE] 9 The display reads 256800000

Now let's try to convert 7.489×10^{-13} to decimal form.

Enter 7.489 [EXP] 13 [±]
The display reads 7.489 $^{-13}$

Press [MODE] 9 The display still reads 7.489 $^{-13}$

Nothing happens, because an exponential notation number with a –13 exponent has too many decimal places to fit into the 10 spaces on the calculator. A number in exponential notation must have an exponent of 9 or less, either plus or minus, to fit into the ten decimal places on a calculator. That is, a number must be between nine billion and nine-billionth.

Now let's convert decimal numbers to exponential notation. Suppose you got 0.00489 as an answer and want to change it to exponential notation.

Enter 0.00489 The display reads 0. 00489

Press [MODE] 8 8
The display reads 4. 8900000 $^{-03}$

Convert 2,090,000 to exponential notation.

Enter 2,090,000 The display reads 2090000

Press [MODE] 8 8 The display reads 2. 090000 ^{06}

Press [MODE] 8 8 to put the calculator in exponential notation mode.

Note: If you press [MODE] 8 8 to convert an answer to exponential notation, the calculator remains in that mode until you press MODE 9 to change to normal decimal mode. If you enter a decimal number, it will change to exponential notation as soon as you press another function key such as [x] or [÷].

Likewise, if you are in decimal mode when you enter a number in exponential notation, it will change to decimal form as soon as you press another key (as long as the decimal form will fit in the display). In either of these cases the number is not affected, only the form of the number.

Let's look at some examples. Press [MODE] 8 8, this puts the calculator in exponential notation mode;

Enter 0.0059 The display reads 0. 0059

Press [×] or [÷]
The display changes to 5. 9000000 $^{-03}$

Press [C] to clear the display. Now press [MODE] 9 to return your calculator to normal decimal mode.

Now enter 0.0059	The display reads 0.0059
Now press [×] or [÷]	The display still reads 0.0059

1.11 Does the Answer Make Sense?

A calculator is a very handy instrument. It will make computations quick and easy, but it is only as good as the numbers you enter. Computer programmers have an acronym for this: GIGO. Garbage in, garbage out. If you are in a hurry on a test, it is easy to push the wrong button and not notice it. Let's look at some calculations and see if we can tell if the answer makes sense.

$$\frac{9.2\times10^{8}}{5.5\times10^{5}}=1.7\times10^{3}$$

We can check the answer to see if it's in the right ballpark by looking at the exponents: $10^8 + 10^5 = 10^3$. Also notice that 9.2 is a little less than twice 5.5, so the ratio of 1.7 is reasonable. Notice that the numerator is larger than the denominator so the answer has a positive exponent.

Let's look at another example:

$$\frac{7.9\times10^{-4}}{6.9\times10^{-9}}=1.1\times10^{5}$$

We look at the exponents: $10^{-4} \div 10^{-9} = 10^{-4-(-9)} = 10^{-4+9} = 10^5$. It checks out. In this case, we divided a larger number by a smaller number and we get a number greater than one. Also notice that 7.9 is just a little larger than 6.9, so a ratio of 1.1 makes sense.

If we had hit the [×] button instead of the [÷] button by mistake, we would have gotten the wrong answer 5.5×10^{-12}. By quickly checking the exponents, we can spot this kind of error. The few seconds you spend checking your answers is time well spent.

Remember that studies show that most students can raise their grades one or even two letters by doing nothing more than being neat and checking their work.

Look at the following problem and the incorrect answer given.

$$\frac{7.50 \times 10^9}{\left(6.5 \times 10^7\right) \times \left(3.5 \times 10^3\right)} = 4.0 \times 10^7 \text{ ???}$$

Do a quick check of the exponents. Did your check tell you that the exponent in the answer should be around 10 and not 10^7? ($10^9 \div 10^5 \div 10^3 = 10^{(9-5-3)} = 10^1$)

Can you figure out what the student did wrong to arrive at the incorrect answer? The student carried out the calculation *incorrectly* as:

7.5 [EXP] 9 ÷ 6.5 [EXP] 5 × 3.5 [EXP] 3 [=]

instead of the correct key sequence:

7.5 [EXP] 9 ÷ 6.5 [EXP] 5 ÷ 3.5 [EXP] 3 =

Remember that on previous pages, we noted that all numbers in the denominator must be preceded by ÷ even if a × sign appears in front of the number. This student got the wrong answer because she forgot this rule.

On multiple–choice tests, the wrong answer you're likely to get is often one of the choices. Let's look at one last example.

$$12 \times \frac{1}{1{,}000} = 12{,}000 \text{ ???}$$

Here the student probably pressed [×] instead of [÷]. Since 1,000 is in the denominator, it will make 12 one thousand times smaller instead of one thousand times larger.

1.12 Practice Problems

Use your calculator. Round all answers to the appropriate number of significant figures. Answers are on the next page.

1. $\dfrac{4.807 \times 1.85}{0.467 \times 5.00} =$

2. $\dfrac{96.8}{15.2 \times 0.0120} =$

3. $10.7 \times \dfrac{1.2}{3.8} \times \dfrac{14.2}{5.01} =$

4. $\left(4.16 \times 10^{5}\right)\left(5.08 \times 10^{-8}\right) =$

5. $\dfrac{7.00 \times 10^{-6}}{5.871 \times 10^{4}} =$

6. $\dfrac{5.68 \times 10^{3}}{5.02 \times 10^{2}} =$

7. $\left(10^{5}\right)\left(10^{-8}\right)\left(10^{3}\right) =$

8. $\dfrac{10^{-14}}{10^{-9}} =$

9. $\left(3.15 \times 10^{18}\right) \times \left(\dfrac{342}{6.02 \times 10^{23}}\right) =$

10. $\left(1.00 \times 10^{15}\right) \times \left(\dfrac{1.66 \times 10^{-24}}{12.0}\right) =$

1.13 Answers to Practice Problems

1. 4.807 [×] 1.85 [÷] 0.467 [÷] 5 [=] 3.808543897

Rounded to 3 sig figs, the answer is 3.81

2. 96.8 [÷] 15.2 [÷] 0.012 [=] 530.7017544

Rounded to 3 sig figs, the answer is 531

3. 10.7 [×] 1.2 [÷] 3.8 [×] 14.2 [÷] 5.01 [=] 9.577056413

Rounded to 2 sig figs, the answer is 9.6

4. 4.16 [EXP] 5 [×] 5.08 [EXP] 8 [±] [=] 0.0211328

Rounded to 3 sig figs, the answer is 0.0211

In exponential notation, the answer is 2.11×10^{-2}

5. 7 [EXP] 6 [±] [÷] 5.871 [EXP] 4 [=] 1.1923011^{-10}

Rounded to 3 sig figs, the answer is 1.19×10^{-10}

6. 5.68 [EXP] 3 [÷] 5.02 [EXP] 2 [=] 11.31474104

Rounded to 3 sig figs, the answer is 11.3

In exponential notation, the answer is 1.13×10^{1}

7. 1 [EXP] 5 [×] 1 [EXP] 8 [±] [×] 1 [EXP] 3 [=] 1

The answer is simply 1

8. 1 [EXP] 14 [±] [÷] 1 [EXP] 9 [±] [=] 0.00001

The answer in exponential notation is 1×10^{-5}

9. 3.15 [EXP] 18 [×] 342 [÷] 6.02 [EXP] 23 [=] 1.7895349^{-03}

Rounded to 3 sig figs, the answer is 1.79×10^{-3}

The decimal form is 0.00179

10. 1 [EXP] 15 [×] 1.66 [EXP] 24 [±] [÷] 12 [=] 1.38333^{-10}

Rounded to 3 sig figs, the answer is 1.38×10^{-10}

Section

2 Skills For Chapter 1: Chemistry: Matter and Measurement

2.1 Significant Figures

Our calculators normally display six to nine digits when performing arithmetic functions. While it may seem advantageous to have such large numbers when performing chemical computations, not all of this information may be significant because of error associated with the data.

To put this into context, consider the data that you record in the laboratory. Every measurement that is made is subject to error. The level of this error (or certainty) depends on the instruments used in the measurement and the skill of the person performing the operations. Even if we can eliminate the systematic errors (*i.e.*, miscalibration of the pan balance) we will still encounter random errors in the laboratory. These random errors will determine the accuracy of our measurement.

Rather than stating the error with each number a chemist may use in a calculation, the practice of using "significant figures" in calculations is employed. The significant figures of a number can be thought of as those digits in the number that do not change when the uncertainty is factored in. (The rules used for determining the number of significant figures in an expressed value are listed in Section 1.4 of your textbook.) As an example of the effect of uncertainty on the number of significant figures, consider a food sample, which was found to have 1.33 mg of Ca per serving with an error of 0.1 mg for the determination. Based on the uncertainty we should only say that the sample contains 1.3 mg of Ca. The number has only two significant figures. It is of no importance to attempt to convey that the calculation indicated a second 3 in the decimal. The error made this digit "insignificant."

You will be asked in this course to use significant figures in the estimation of answers to your calculations. As an example, consider the problem of determining the circumference of a cylinder. With a simple ruler you could determine the diameter to be 2.5 inches. From the relationship of $C = \pi \times d$, you could use your calculator to determine the circumference, C. The display of the calculator would read (7.853981...). What value do you report? What is significant?

The basic rule is that you report your answer using the least number of significant figures. In the multiplicative operation above, the diameter was only reported to two significant figures. The answer should be reported as 7.9 in.

Rules for addition and subtraction are slightly different. Consider adding a 0.001 g weight to an object resting on a pan balance and indicating a value of 23.2 g in mass. The addition of 0.001 to 23.2 would yield a theoretical answer of 23.201 g. However, if the pan balance used had an error of 0.1 g, the addition of 0.001 g to this weight would be insignificant. The uncertainty will limit our ability to express accurately the sum of the combined weights. Sample Exercises 1.4–1.6 show worked problems on significant figures. Problems 49–58 of Chapter 1 deal with significant figures.

2.2 Exponential Notation

Exponential notation is a method used to express very large and very small numbers on your calculator. These numbers are expressed by multiplying the significant portion of the number by a multiplier based on 10^x.

To use the exponential notation feature on your calculator, you will use a [sci], [exp] or [EE] function. Consider the process of entering 1.6×10^{-17}. First enter the significant portion of 1.6 into the display. Next press the [sci] or equivalent key and enter the number 17. You will need to change the sign of the power of 17 (which should appear in the right hand portion of the display), so press the [+/-] key to cause the power to read –17. Depending on the type of calculator you have, you may need to press the [=] key to end the sequence.

You will also be able on your calculator to convert from regular notation to exponential notation. If the number 1,000,000 was in the display of your calculator, you could depress the [sci] then [=] keys to cause the number to appear in exponential notation (1.000×10^6) on the display.

An example of exponential notation concerns a problem that you will encounter in Chapter 7 of your textbook where you will determine the energy of a photon. If a HeNe laser has a wavelength of 632.8 nm, what is the energy in J of a single photon? The answer is calculated using $E = hc/\lambda$, where h and c are constants and λ is the wavelength in meters. We can quickly estimate the energy by expressing all numbers in exponential notation and setting up the problem. Using this, we can write:

$$h = 6.6262 \times 10^{-34}\ \text{J·s}$$
$$c = 2.998 \times 10^{8}\ \text{m/s}$$
$$\lambda = 632.8\ \text{nm or } 632.8 \times 10^{-9}\text{m}$$

or 6.328×10^{-7} m

The problem can be set up to solve for E as:

$$E = (6.626 \times 10^{-34}\text{J·s}) \times \frac{2.998 \times 10^{8}\, m/s}{6.328 \times 10^{-7}\, m}$$

The calculator would allow us to find that there are 3.139×10^{-19} J/photon, based on 4 significant figures.

More information on exponential notation can be found in Appendix A of your textbook. Problems 53 and 54 of Chapter 1 deal specifically with the expression of numbers using this format. Also, see Sections 1.8–1.10 of this toolkit for more discussion on exponential notation.

2.3 Mean and Standard Deviation

If your calculator has a statistical analysis package, you may be able to calculate the mean ($\bar{x}$) and standard deviation (s) of a data set.

These values are helpful in determining the accuracy and precision of your measurements.

Let us say that you have titrated five equal volumes of an unknown acid with a standard base. You have dispensed the following volume of titrant in the 5 runs: 43.80 mL, 43.30 mL, 44.10 mL, 43.90 mL and 43.70 mL. We would first enter the five values into the memory of the calculator. After final entry, we would depress the button to give us the mean, $\bar{x}$. The value of 43.76 should appear.

Next, we would calculate the standard deviation, s, by pressing the appropriate buttons (s, s_x, or σ). A value of 0.30 should appear. Based on your titration skills you could say you dispensed 43.76 +/- 0.30 mL per titration in this data set. Considering the error in the data (0.30 mL) expressing a value beyond the tenths of mL (0.1 mL) would not be significant. The correct answer should be 43.8 ± 0.3 mL, where only three significant figures are used.

Section 3 Skills for Chapter 3: Stoichiometry: Chemical Calculations

Goals: to calculate yields, concentrations, dilutions, and titrations
Skills: multiplication and division

3.1 Yields

Part of your studies dealing with chemicals and their reactivity will be to determine the yield of a reaction. From your readings concerning stoichiometry, if we write the equation,

$$A + B \rightarrow C$$

we understand if 1 mol of A and 1 mol of B are allowed to react under favorable conditions, 1 mol of C should be produced. The *theoretical yield* of the reaction is 1 mol of C. This value could be expressed in grams instead of moles by multiplying the theoretical yield of 1 mol by the molar mass of the compound C.

Note that yield depends on the stoichiometry of the reaction. If the reaction were instead,

$$2\,A \rightarrow B$$

and 1 mol of A was placed in the reaction vessel, the theoretical yield would be 0.5 mol of B. Using the problem-solving method (Sections 1.5 and 1.6 of your text) the solution could be found.

$$x \text{ mol} = 1 \text{ mol A} \times \frac{1 \text{ mol B}}{2 \text{ mol A}}$$

$$x \text{ mol} = 0.5 \text{ mol B}$$

In practice, the mol quantity of product that is obtained from the reaction is less than the theoretical value. The amount of product that is obtained is termed the *actual yield.*

A method in which the efficiency of a reaction is gauged is the calculation of the *percent yield.* It is defined as

$$\textit{percent yield} = \left(\frac{\textit{actual yield}}{\textit{theoretical yield}}\right) \times 100\%$$

For the first reaction, if we had obtained 0.8 mol of C, the % yield would be

$$\textit{percent yield} = \left(\frac{0.8 \text{ mol}}{1 \text{ mol}}\right) \times 100\%$$

$$\textit{percent yield} = 80\%$$

The calculation of percent yield can also be performed using the grams of actual product obtained as compared to the theoretical gram yield.

In many cases, the scientist may wish to predict the actual yield for a well-characterized reaction. For instance, if the percent yield of a synthetic reaction is 90%, the actual yield may be found by multiplying the percent yield by the theoretical yield. (Remember to divide the percent yield by 100% in order to remove the % unit from the expression.) Examples 3.21 and 3.22 use variations of this approach to determine the mass of a product, based upon the percent yield of the reaction. Additional problems are 81–84 of Chapter 3.

3.2 Molarity

The molarity, M, is the term that expresses the mol quantity of a dissolved substance (solute) per liter of solution. A solution that is 1.00 M NaCl contains 1.00 mol of NaCl per 1.00 L of solution. This same solution contains 58.4 g of NaCl per 1.00 L of solution.

Calculations that involve molarity will usually begin with statement of the gram quantities of solute dissolved in a quantity (mL or L) of solution. By definition

$$M = \frac{\text{mol of solute}}{\text{L of solution}} = \frac{\left(\dfrac{\text{g of solute}}{\text{MW of solute}}\right)}{\text{L of solution}}$$

For instance, to calculate the molarity of a solution that is 10.0 g of NaCl in 400 mL of water, we would write:

$$M = \frac{\left(\dfrac{10.0\ \text{g}}{58.4\ \text{g / mol}}\right)}{0.400\ \text{L}}$$

We will first solve the numerator of the problem for the mol amount of NaCl present in the solution. If we divide 10.0 g by 58.4 g/mol, we should obtain a value of 0.171 mol NaCl. Substituting into the equation

$$M = \frac{0.171\ \text{mol}}{0.400\ \text{L}}$$

$$M = 0.428\ \text{mol NaCl / L of solution}$$

Note: It is important to express the volume of solution only in L.

See Example 3.23–3.25 and Problems 85 and 86 in Chapter 3 of your textbook for similar problems.

You may find it necessary in your work to determine what amount of solute, expressed in grams, would result in a certain molarity of solution. Consider what gram amount of NaCl is necessary to create a 0.60 M solution that is 750 mL in volume. First, let's fill the given information into the molarity expression.

$$0.60\ M = \frac{\left(\dfrac{x\ \text{g}}{58.4\ \text{g / mol}}\right)}{0.750\ \text{L}}$$

To solve for x, we must cross multiply several factors in order to arrange the equation for x.

$$(0.60\text{ M})\ (0.750\text{ L}) = \frac{x\text{ g}}{58.4\text{ g/mol}}$$

$$(0.60\text{ mol/L})\ (0.750\text{ L})\ (58.4\text{ g/mol}) = x\text{ g}$$

$$x\text{ g} = 26.3\text{ g of NaCl}$$

Note that all units on the left hand side of the equation reduced to g (remember that M is in mol/L). By taking time with your cross multiplication step, the problem can be solved. This example is similar to Problems 87 and 88 in Chapter 3 of your textbook.

3.3 Dilution

Dilution involves reducing the concentration of a solution by adding more solvent. Typically, a volume of solution is taken and added to a volume of solvent. Consider the dilution of a 3.00 M NaCl solution by pipetting 10.0 mL into a 1.000 L flask and filling the flask to the calibration mark with pure water. Here we are adding roughly 990 mL of water to the solution.

We can calculate the concentration of a diluted solution using the definition of molarity and the general problem solving method of Sections 1.5 and 1.6 of your text. Alternatively, we can use the equation

$$M_f = M_i \times \frac{V_i}{V_f}$$

where M_f is the final concentration of the diluted solution, M_i is the initial concentration of the concentrated solution, V_f is the final concentration of the diluted solution, and V_i is the initial concentration of the solution. For our problem, M_i = 3.00 M NaCl, V_i = 10.0 mL, and V_f = 1.000 L (or 1,000 mL). Substituting these numbers into the equation, we write

$$M_f = 3.00\text{ M} \times \frac{10.0\text{ mL}}{1000\text{ mL}}$$

$$M_f = 0.0300 \text{ M}$$

Note that the volume units in a dilution calculation must be the same no matter how they are stated in the problem.

Use the problem solving method of analysis to check your units before reporting your answers. See Problems 93 and 94 of Chapter 3 for more practice.

The dilution equation can also be used to solve for V_i, if the other variables of the equation are known. For instance, what volume (V_i) of a 1.00 M solution of NaOH (M_i) must be added to a beaker and diluted to 500.0 mL (V_f) in order to produce a diluted solution of 0.2500 M NaOH (M_f). Rearrangement of the dilution equation by cross-multiplication and division yields

$$V_i = V_f \times \frac{M_f}{M_i}$$

from which V_i can be found. Example 3.26 and Problems 95–98 are similar to this worked example.

3.4 Titration

Another calculation for solutions that involves cross multiplication and division is titration. As discussed in your textbook, this process involves reacting a standard solution with an unknown solution for the purpose of determining the concentration of the unknown solution. In the lab, the commonly used titrations involve acid-base reactions, precipitation reactions, redox reactions and complexation reactions.

Before any titration can be carried out, the solution stoichiometry must be known. Specifically, we must know how many moles of the titrant (standard solution) will react with the unknown. Using the simple acid-base reaction listed below, the mathematics for a titration can be set up.

$$HCl(aq) + NaOH(aq) \rightarrow H_2O(l) + NaCl(aq)$$

This reaction has a stoichiometry of 1 mol of acid to 1 mol of base. We can then write:

$$mol_{acid} = mol_{base}$$

Substituting the expression of mol = M × V, we obtain:

$$M_{acid} \times V_{acid} = M_{base} \times V_{base}$$

Consider a titration of an unknown NaOH solution with a standardized HCl solution. To 50.0 mL of the unknown contained in a flask, approximately 33.8 mL of a 0.175 M standardized HCl solution was added to the flask to reach the equivalence point. From this information, we want to find out the molarity of the unknown solution.

If we rearrange the above equation by cross multiplication and division, we can solve for the molarity of the unknown base, M_{base}. Doing so yields:

$$M_{base} = \frac{M_{acid} \times V_{acid}}{V_{base}}$$

Putting in the data, we find:

$$M_{base} = \frac{0.175 \text{ mol/L} \times 0.0338 \text{ L}}{0.0500 \text{ L}}$$

(Note: volumes are in L.)

$$M_{base} = 0.118 \text{ M}$$

Please note that these reactions noted above possess a solution stoichiometry of 1:1. For different solution stoichiometry, see Example 3.27 of your text. In this example the base is $CaCO_3$. It consumes 2 moles of HCl in the titration reaction (the relationship is 2 mol_{acid} = 1 mol_{base}). Problem 105 is similar to this example.

Section

4 Skills for Chapter 5: Gases

Goals: to perform calculations with gas laws, partial pressure, and effusion equations

Skills: cross multiplication, division, and roots

4.1 Gas Laws

In this section of your studies, you will be exploring relationships between pressure, P, volume, V, temperature, T, and mol quantities of gases, n. Your text will introduce you to Boyle's law, Charles's law, Avogadro's law, the Ideal Gas law, Dalton's Law and Graham's law.

You must exercise care in two areas to avoid mistakes with these equations. The first area is temperature: all temperatures must be expressed in Kelvin (K) even if the data is given to you in °C. Remember, the relationship is K = °C + 273.15. The second area is to rearrange the equation so that the left hand side only contains the term that you are seeking. All the given data would then appear on the right hand side of the equation.

Consider the use of Charles's law in finding the volume a gas in a balloon would occupy if it was heated from 25°C to 75°C. (This law is discussed in Section 5.5 of your textbook.) The volume of the gas at 25°C is 1.00 L. First, let's set up the equation:

$$\frac{V_{\text{initial}}}{T_{\text{initial}}} = \frac{V_{\text{final}}}{T_{\text{final}}}$$

From the above information we can write:

$$V_{\text{initial}} = 1.00\text{ L}$$
$$T_{\text{initial}} = 25°\text{C, which is } 298\text{ K}$$
$$V_{\text{final}} = ?$$
$$T_{\text{final}} = 75°\text{C, which is } 348\text{ K}$$

To solve for V_{final}, we write:

$$\frac{V_{\text{initial}}\,T_{\text{final}}}{T_{\text{initial}}} = V_{\text{final}} \quad \text{(cross multiplying } T_{\text{final}}\text{)}$$

$$V_{\text{final}} = \frac{V_{\text{initial}}\,T_{\text{final}}}{T_{\text{initial}}} \quad \text{(switching sides)}$$

By plugging in the data, we obtain:

$$V_{\text{final}} = \frac{1.00\text{ L} \times 348\text{ K}}{298\text{ K}}$$

$$V_{\text{final}} = 1.17\text{ L}$$

See Examples 5.6 and 5.7 as well as Problems 37–40 of Chapter 5 for more work with Charles's Law.

When dealing with Boyle's law (Examples 5.4, 5.5 and Problems 31–36), Charles's law, Avogadro's law (Example 5.8 and Problems 41–46), or the Combined Gas law (Example 5.9 and Problems 47–52) you will often find much information in the problem. Follow these steps to solve such problems:

1) write the gas law equation to be used.
2) sort the given data (as set 1 or set 2).
3) rearrange the equation to solve for the unknown term.
4) enter the data and solve.

4.2 Ideal Gas Law

Another gas law that you will use is the Ideal Gas Law, $PV = nRT$. It is described in Section 5.8 of your textbook. Here n is the number of moles of the gas and R is the gas constant (0.0821 L atm mol^{-1} K^{-1}). This gas law will allow you to work any of the previous problems that relate the P, V, and T of a gas to the mole quantity of the gas. For instance, what volume would 1.00 g of CO_2 occupy at 1.00 atm of pressure and a temperature of 23°C? Begin this problem by writing the equation.

$$PV = nRT$$

We were not given n in the original problem, but we can calculate it by using the equation n = *g/molar mass.* Here,

$$n = \frac{1.00\text{ g}}{44.0\text{ g / mol }CO_2}$$

and is equal to 0.0227 mol. With this information we can write:

P = 1.00 atm
V = ?
n = 0.0227 mol
R = 0.0821 (0.0821 L atm mol^{-1} K^{-1})
T = 23°C, which must be expressed as 296 K

Rearranging the equation to solve for V, we write:

$$V = \frac{nRT}{P}$$

Plugging in the data:

$$V = \frac{0.0227\text{ mol} \times 0.0821\text{ (L atm mol}^{-1}\text{ K}^{-1}) \times 296\text{ K}}{1.00\text{ atm}}$$

V = 0.551 L

More examples using this equation are found in Examples 5.10–5.11 and Problems 53–58 of your text. The Ideal-Gas law is further used for the calculation of molecular weights of compounds in

Examples 5.12–5.13 and Problems 59–62, and for gas densities in Examples 5.14–5.15 and Problems 63–70.

4.3 Dalton's Law

The pressure of a gaseous mixture is dependent on the partial pressure exerted by each gas. Dalton's law of partial pressures describes this relationship and is expressed as P_1 = *mole fraction of compound*$_1$ × P_{total}, where P_1 is the pressure of gas 1and P_{total} is the total pressure of the gas. To illustrate this law, consider a container where 1.0 mol of gas 1 is mixed with 3.0 moles of gas 2. The mole fraction of gas 1 is calculated as follows:

$$\text{mole fraction of compound}_1 = \frac{\text{1.0 mole of gas 1}}{\text{4.0 moles of gas total}}$$

$$\text{mole fraction of compound}_1 = 0.25$$

similarly,

$$\text{mole fraction of compound}_2 = \frac{\text{3.0 moles of gas 2}}{\text{4.0 moles of gas total}}$$

$$\text{mole fraction of compound}_2 = 0.75$$

IMPORTANT: The sums of all the mole fractions should always add up to 1. Since the mole fraction is a simple ratio, it has no units.

Now, to work the problem, let's assume the pressure in the container was measured to be 2.00 atm. What is the partial pressure of gas 1? The equation is:

$$P_1 = \textit{mole fraction of compound}_1 \times P_{total}$$

The terms to use in the equation are:

mole fraction of compound$_1$ = 0.25 (calculated above)

$$P_{total} = 2.00 \text{ atm} \qquad \text{(given in the problem)}$$

Now by substituting into the equation we obtain,

$$P_1 = 0.25 \times 2.00 \text{ atm}$$
$$P_1 = 0.50 \text{ atm}$$

By similar reasoning we could determine that $P_2 = 1.50$ atm. Also since the gas mixture consisted of only 2 gases, if the pressure of gas 1 is 0.50 atm, and the total pressure of *both* gases is 2.00 atm, by difference we can find the pressure of gas 2 to be 1.50 atm. See Examples 5.18–5.20 and Problems 81–84 for more work with Dalton's Law.

A special extension of Dalton's Law is addressed in Example 5.21 and Problems 85-88 of Chapter 5. Here Dalton's Law is used to correct the pressure reading of a gas that is collected over water. The use of the law allows you to *remove* the P_{H_2O} from P_{total} based upon the temperature of the water.

4.4 Graham's Law

This law relates the effusion rate of a gas to the inverse square root of the molar mass of the gas. The theory behind this equation is found in Section 5.11 of your textbook. You will normally use this law in comparison as:

$$\frac{\text{rate}_1}{\text{rate}_2} = \sqrt{\frac{M_2}{M_1}}$$

With this relationship, we must be careful with the square root sign. To demonstrate how to solve a problem with this equation let's use the following data (and for the moment forgo the units): $\text{rate}_1 = 2$, $\text{rate}_2 = 1$, $M_1 = ?$, and $M_2 = 100$. We must rearrange for M_1. To do so, first let's square both sides of the equation. This process will remove the square root sign.
(Remember $\sqrt{x^2} = x$)

$$\left(\frac{\text{rate}_1}{\text{rate}_2}\right)^2 = \frac{M_2}{M_1}$$

Through cross multiplication, we can write

$$M_1\left(\frac{\text{rate}_1}{\text{rate}_2}\right)^2 = M_2$$

And finally through rearrangement, the expression can be written for the unknown parameter.

$$M_1 = M_2\left(\frac{\text{rate}_2}{\text{rate}_1}\right)^2$$

Using the data given in the problem.

$$\text{rate}_1 = 2$$
$$\text{rate}_2 = 1$$
$$M_1 = ?$$
$$M_1 = 100 \times \left(\frac{1^2}{2^2}\right)$$
$$M_1 = \frac{100}{4}$$
$$M_1 = 25$$

Whenever we encounter equations with square roots, we must deal with the square root and rearrange the equation before we solve for the unknown.

Another example of a calculation using Graham's law is shown in Examples 5.23 and 5.24 of your textbook. Problems 93–94 also relate to Graham's law.

Section

5 Skills for Chapter 6: Thermochemistry

Goals: to calculate heat evolved in calorimetry and heats of reaction
Skills: addition, multiplication and division

5.1 Calorimetry

Calculations in calorimetry involve measuring the change in the temperature of a chemical reaction. This is normally accomplished by noting the temperature of the reaction mixture before and after the chemical reaction has taken place. In many cases, the reaction is contained in a special measuring device called a *calorimeter*. For a reaction where dilute solutions of strong acid and base are mixed together in a calorimeter, the heat evolved in the reaction is:

$$q = C \times \Delta T$$
$$q = (\text{mass}) \times (\text{specific heat}) \times \Delta T$$

This expression is found on Section 6.5 of your textbook. Here C is the heat capacity of a system and ΔT is the change in the temperature of the system. Note that C is equal to the mass × specific heat. The mass for the mixture is determined by weighing or measuring the volume of liquid in the calorimeter and multiplying this volume by the density of the solution. The specific heat is given in a table. The ΔT is calculated from the temperature readings. The ΔT term is defined as the final temperature minus the initial temperature. A reaction where heat is evolved (exothermic) will have a positive ΔT while a reaction where heat is absorbed from the environment (endothermic) will have a negative ΔT.

Let us say that when the acid and base were mixed in the calorimeter, the temperature of the solution changed from 23.5°C to

25.9°C. The final volume of the mixture was 50.0 mL. Since the density of the solution (being mostly water) is 1.00 g/mL, the aqueous solution weighs 50.0 g. The specific heat of water is 4.18 J/°C g. Hence, the heat capacity of the solution is 209 J/°C. The ΔT is calculated as 25.9 – 23.5°C = 2.4°C. Placing these values into the above equation yields:

$$q = (209\ \text{J/°C}) \times (2.4\text{°C})$$
$$q = 502\ \text{J}$$

See Example 6.6 and Problems 43 and 44 of your textbook for more problems in calorimetry.

Several variations on these examples appear in your text. One involves finding out how much heat is needed to raise the temperature of a certain mass of water (Example 6.7, Problem 45 and 46). One involves the prediction of the final temperature of a substance based upon the mass of the substance and q (shown in Example 6.8 and Problems 47 and 48). Another concerns the determination of the mass of the substance based upon q (Problem 52). The calculation of the specific heat is the focus of Problems 49 and 50.

A point to remember is that any expression that involves reactions will deal with the difference in the final state from the initial state. For any calculation using Δ (*i.e.*, ΔT, $\Delta H°$) make a habit of thinking that Δ = (final state – initial state). Using this approach you won't get the sign of the Δ wrong.

5.2 Standard Enthalpy of Reaction

The determination of the heat of reaction involves the use of the standard enthalpy of formation found in Table 6.2 and Appendix C.1 of your textbook. The enthalpy change for a reaction (ΔH) is the difference in the sums of the standard enthalpy of formation ($\Delta H°_f$) of the products and the reactants in a reaction. In the calculation, we must take care to observe the sign (positive or negative) associated with the standard enthalpy of formation in order to obtain the correct answer. Consider the combustion of methane gas:

$$CH_4(g) + 2\,O_2(g) \rightarrow CO_2(g) + 2\,H_2O(l)$$

The calculation of ΔH in kJ/mol for this reaction is:

$$\Delta H = [-393.5 + 2(-285.8)] - [-74.8 + 0]$$

These values are found in Table 6.2 and in the appendix. (*Note*: O_2 has a $\Delta H°_f$ of zero.) To begin solving, first determine the number within each set of brackets.

$$\Delta H = [-965.1] - [-74.8]$$
$$\Delta H = -965.1 + 74.8$$
$$\Delta H = -890.3 \text{ kJ/mol}$$

Further examples of these calculations are shown in Examples 6.15–6.18, and Problems 71–78.

In addition to observing the sign of the $\Delta H°_f$, pay close attention to the stoichiometry and the state of the products and reactants. Different states of a species (gas or liquid) will have different $\Delta H°_f$ values. Examine Table 6.2 of your textbook for values assigned to the different states of water.

Section

6 Skills for Chapter 7: Atomic Structure

Goals: calculation of frequency and wavelength of photons, use of the Bohr Energy-Level equation

Skills: multiplication and division

6.1 Radiant Energy

In your study of radiant energy, you will investigate the relationship of the wavelength and frequency of photons. The basic equation is:

$$c = \nu \times \lambda$$

where ν is the frequency of the photon (expressed in Hertz) and λ is the wavelength of the photon (expressed in m). The speed of light, c, is a constant and is approximately 3.00×10^8 m/s. Care must be taken in this calculation to keep all values in the correct units. For instance, λ of a sodium vapor lamp is 589 nm. To determine ν we can use the above equation. However, all values involving a measure of length in this calculation need to be expressed in the same units. The most convenient unit is meters for this problem. The wavelength is in nm, so the value would be expressed as 589×10^{-9} m, or better yet 5.89×10^{-7} m. By rearranging the above equation to solve for ν, we write:

$$\nu = c / \lambda$$
$$\nu = (3.00 \times 10^{8} \text{ m/sec}) / 5.89 \times 10^{-7} \text{ m}$$
$$\nu = 5.09 \times 10^{14} \text{/sec or Hz}$$

The solution to a similar problem for an X-ray source is shown in Example 7.1 of your text. Exercise 7.1b addresses the determination of λ when given ν. Additional exercises are Problems 31–36.

This conversion of wavelength to frequency is very useful in dealing with problems in the quantum theory section where the energy of a photon is related to the frequency of the electromagnetic radiation using $\Delta E = h\nu$. Example 7.3 and 7.4 shows such a calculation. Problems 37–40 are similar. Note that the photoelectric effect (Problems 41–44) also makes use of this relationship.

6.2 Bohr Energy-Level Equation

The Bohr Energy-Level equation makes use of quantum theory and relates the energy of a photon that corresponds to an electronic transition within the orbitals of an H atom. This relationship is also used to match the emission spectra of the H atoms with electronic transitions that may occur between the orbitals. The equation is

$$\Delta E_{level} = B\left(\frac{1}{n_i^2} - \frac{1}{n_f^2}\right)$$

The n terms are integer values where n_i is the principal quantum number of the initial and n_f is the principal quantum number of the final orbital of the electron. The term B is a constant with a value of 2.179×10^{-18} J.

Take care when calculating the difference in the inverse squares of the n_i and n_f values. If $n_i = 2$ and $n_f = 4$ the calculation of the difference in the inverse square is:

$$x = \left(\frac{1}{2^2} - \frac{1}{4^2}\right)$$

$$x = \left(\frac{1}{4} - \frac{1}{16}\right)$$

$$x = \frac{^-3}{16}$$

A common mistake that occurs in this calculation is to write the difference of the inverse squares of *n* as:

$$\frac{1}{x} = \left(\frac{1}{2^2 - 4^2}\right) \quad \textit{Wrong}$$

or

$$\frac{1}{x} = \left(\frac{1}{2 - 4}\right)^2 \quad \textit{Wrong}$$

Both of these expressions will produce *incorrect* answers. You must square the *n* value with the [x^2] key, invert the value with the [1/*x*] key and then take the difference in the values. Then you may solve for *x*.

An example of the calculation of the energy emitted by an H atom when an electron falls from $n_i = 5$ to $n_f = 3$, is found in Example 7.6. A similar calculation is used in Problem 45 of your textbook. Problem 46 deals with the absorption of energy by hydrogen, which is the reverse of emission.

The wavelength of a photon emitted from an H atom can be found by determining ΔE_{level} and then entering this value into the equation $\Delta E_{level} = hc/\lambda$. See Example 7.7 and Problems 47 and 48.

6.3 Wavelike Properties of Matter

Another useful relationship presented in Chapter 7 is the De Broglie equation, which describes the λ of a particle of known mass *m* and traveling at a velocity *v*. The equation is found in Section 7.8 of your text and appears as:

$$\lambda = \frac{h}{m\,v}$$

Here h is Planck's constant (6.626×10^{-34} J s). Solutions for λ or v with this equation involve simple multiplication and division. Example 7.9 of your text shows the use of this equation for determining λ of the matter waves of traveling particles.

Care must be taken when using this equation to use the correct units for m and v. Note that 1 J = 1 kg m^2 s^{-2}. All m values must be expressed in kg and v values in m/s. Problems 53–56 make use of this equation.

Section

7

Skills for Chapter 9: Chemical Bonds

Goals: to calculate heats of reaction based on bond energies
Skills: addition and subtraction

7.1 Bond Dissociation Energies

The enthalpy change in a chemical reaction can be calculated by assessing the number and types of chemical bonds broken and formed. The breaking of chemical bonds requires energy from the surroundings while the formation of chemical bonds releases energy back to the surroundings. By tabulating the energy input and output of a reaction, the $\Delta H°$ of a reaction can be determined.

In Section 9.10 of your text, a simple formula is presented for determining ΔH based on bond dissociation energies (D). The D values are found in Table 9.1. Example 9.14 shows the use of this equation to find $\Delta H°$ for the production of N_2H_4 from N_2 and H_2. By knowing the chemical form of the products and reactants and the types of bonds involved in the chemical reaction, the D values can be entered into the equation to find ΔH. Problems 69–72 involve this principle. A variation on this concept is the determination of a D value for a chemical reaction when $\Delta H°$ is given. This is demonstrated in Problems 73–76.

Section

8 Skills for Chapter 12: Solutions

Goals: to determine concentrations of solutions and calculate the effect of solution strength on colligative properties

Skills: cross multiplication, division

8.1 Concentration Units

The calculations that you will perform in this chapter involve molarity, percent by mass, percent by volume, mass/volume percent, molality, mole fraction and mole percent. These definitions are outlined in Section 12.2 of your text. Although they may seem similar, they involve very different units. *Molarity* involves *L* of solvent, while *molality* involves *kg* of solvent. You must use the exact units specified in the definition in order to calculate the correct concentration. For instance, if you are told that 1.0 mole of solute was dissolved in 500 mL of water, you would calculate the molarity as 2.0 M (that is, 1.0 mole of solute in 0.500 L of solution, or 2.0 mole of solute in 1.000 L of solution). The use of mL instead of L would give a wrong answer. Review Examples 12.1–12.7 and Problems 21–42 for more examples.

8.2 Colligative Properties

There are several formulas to examine in this chapter. They are:

$P_A = \chi_A P_A$	Raoult's Law (Section 12.6)
$\Delta T_b = K_b m$	Boiling point elevation (Section 12.7)
$\Delta T_f = -K_f m$	Freezing point depression (Section 12.7)
$\pi = \frac{n}{V} RT$	Osmotic Pressure (Section 12.8)

Examples 12.10–12.13 and Problems 53–56 demonstrate the use of Raoult's Law. Examples 12.14, 12.15 and Problems 57–64 deal with freezing point depression and boiling point elevation, while Example 12.16 and Problems 65-70 cover osmotic pressure. Further examples of colligative properties using strong electrolytes, weak electrolytes, and nonelectrolytes are found in Problems 71–78.

You will use the concentration units in the above section to calculate the effect of the solution strength on these colligative properties. The calculations are straightforward: the errors in the computations most often involve the use of the wrong constants (k_f vs. k_b), incorrect units (mL vs. L), or wrong concentration expression. The boiling point and freezing point equations are very similar but use different k values. For freezing point depression problems you can only use the k_f value for the solvent and not the k_b value. For osmotic pressure, the unit in the gas constant dictates the units in the equation. If 0.08206 L atm $K^{-1}mol^{-1}$ is used, the pressures must be in atm and the temperatures in K. Because our bodies are usually at 37°C, we must use 310 K in osmotic pressure problems relating to biochemistry.

Finally, you must remember that there is a difference between m and M. The majority of the solution calculations that you will use in this course will involve molarity, M. The molality definition, m, is used much less frequently and mainly occurs in the freezing point depression and boiling point elevation problems.

Section

9 Skills for Chapter 13: Chemical Kinetics

Goals: to calculate reaction rates, concentrations as a function of time, half-lives, and to examine the relationship of rate, temperature and activation energy

Skills: multiplication, division, logarithms, powers

9.1 Reaction Rates

The reaction rate expresses how the concentration of a product or reactant changes during the course of a reaction. The most convenient way to express this change is to measure the concentration as a function of time. For a simple reaction such as:

$$A \rightarrow B$$

the rate can be expressed as how the concentrations of A or B change during the time period in which the reaction is observed. If we were to plot the concentration of A as function of time, we could generate a plot, such as Figure 13.5 in your text. The rate of change in A is calculated as the change in [A] vs. the change in time:

$$\text{rate of change in A} = \frac{-\Delta[A]}{\Delta t}$$

The negative sign in front of Δ[A] indicates that [A] is diminishing as the reaction proceeds. This makes sense, because A is the reactant. If we wished to determine the rate of this reaction as expressed by B, we would write the rate expression as $\Delta[B]/\Delta t$. Here the sign associated with B is positive; B is appearing in the reaction vessel because it is the product.

You may recognize the reaction rate as the slope of the line in Figure 13.5. If your calculator contains a statistical analysis package, you could enter the data ([A] as *y* and *t* as *x)* and compute the slope of the observed points. As your text mentions, some of these relationships are not linear. Just as the slope changes along the various points of Figure 13.5, so will the reaction rate.

An important point to remember in working with reaction rate problems is the order of the reaction with respect to each reactant. You may, after observing the reaction rates for several different concentrations of reactants, note that a reaction is best described by the rate law:

$$\text{rate of the reaction} = k[A]^2[B]^1$$

The reaction is *second order* with respect to A and *first* order with respect to B. If [A] is doubled, the rate of the reaction should quadruple (*i.e.* $4 = 2^2$). If [B] alone is doubled, the rate of the reaction should only double (because of the linear relationship between rate and concentration for a first order dependency).

If you are asked to determine the rate constant *k* with a given rate and [A] and [B], you should first rearrange the equation and solve for *k*:

$$k = \frac{\text{rate of the reaction}}{[A]^2[B]^1}$$

If [A] = 0.100 M, [B] = 0.200 and rate = 6.0×10^{-4} M/s

$$k = \frac{6.0 \times 10^{-4}\ \text{M/s}}{[0.100\ \text{M}]^2[0.200\ \text{M}]^1}$$

$$k = \frac{6.0 \times 10^{-4}\ \text{M/s}}{0.0100\ \text{M}^2 \times 0.200\ \text{M}}$$

$$k = 0.30\left(\text{M}^{-2}\text{s}^{-1}\right)$$

Example 13.3 is similar to this problem, where the reaction of NO with Cl_2 is studied. For more work, see Problems 23–36. Note the similarity of Example 13.3 with Problems 31 and 32. Problem 31 is a variant of this example in which a reaction having a different stoichiometry is used.

One special case to remember when working with reaction rates is zero order. If a reactant is found to have a zero order influence on the rate of the reaction, it will not influence the rate. For instance, if the rate of the reaction were:

$$\text{rate of the reaction} = k[\text{A}]^1[\text{B}]^0$$

the rate is unaffected by [B]. (Any number raised to the zero power is one. Try it with your [y^x] key using $x = 0$.) Hence, the reaction is only first order with respect to A.

9.2 Concentration and Time

You will work with *first* and *second order* reactions in your study of General Chemistry. Equations have been developed that relate the concentration as a function of time. The *first order* equation involves a logarithmic relationship and is best expressed as:

$$\ln\left(\frac{[\text{A}]_t}{[\text{A}]_0}\right) = -kt \qquad \text{or} \qquad \ln[\text{A}]_t = -kt + \ln[\text{A}]_0$$

where rate = k[A]. This equation is found in Chapter 13.5 of your text.

A common use of the integrated rate equation involves the determination of the concentration of A after a certain time has elapsed in the reaction. To determine this concentration, we must first solve for $\ln[\text{A}]_t$ and them transform this value. Consider a problem where $k = 0.00500/\text{s}$, $[\text{A}]_0 = 1.00$ M, and $t = 100$ s. The solution for $[\text{A}]_t$ begins with:

$$\ln[\text{A}]_t = -(0.00500/\text{s} \times 100\text{s}) + \ln[1.00]$$

$$\ln[A]_t = -(0.500) + 0$$

$$\ln[A]_t = -0.500$$

To solve for $[A]_t$, the *ln* expression must be eliminated. This can be legitimately done if we consider the relationship of *ln(x)* and 10^x. By applying the "inverse-log" (that is, e^x) to the both sides of the equation, the term is eliminated. If you take the *ln* of a number, applying the inverse-log would restore the original number. In essence, e^x will undo what *ln(x)* does. (For base_{10} math, *log(x)* and = 10^x are similarly related.)

Therefore, to solve for $[A]_t$, we will algebraically apply e^x to both sides of the equation. Doing so yields:

$$e^{\ln[A]_t} = e^{(-0.500)} \quad \text{and since} \quad e^{\ln[A]_t} = [A]_t$$

$$[A]_t = e^{(-0.500)}$$

$$[A]_t = 0.606\,M$$

Use this approach to solve the concentrations of first order decay reactions such as Example 13.4 and Problems 37–42.

It is most important that you understand how the *ln* term was eliminated and the A value obtained in the previous problem. Many of the problems that you will encounter in later chemistry chapters will be solved with the "*ln*, inverse-*ln*" relationship.

Another use of the integrated rate law equation for a first order reaction involves the determination of *t* based upon *k*, $[A]_t$, and $[A]_0$. For instance, at what time *t* will A fall from 1.00 M to 0.20 M if the *k* is 0.00500/s. Substitution of these values into the equation yields:

$$\ln\left(\frac{[0.20]_t}{[1.00]_0}\right) = -(0.00500\,/\,s)t$$

$$\ln(0.20) = -(0.00500\,/\,s)t$$

$$-1.609 = -(0.00500\,/\,s)t$$

$$t = \frac{-1.609}{(-0.00500/s)}$$

$$t = 320s$$

Note that this problem is similar to Example 13.4. Also, use this approach in solving Problem 39c.

A common mistake made in working these problems involves the following incorrect assumption:

$$\ln\left(\frac{[A]_t}{[A]_0}\right) = \frac{\ln [A]_t}{\ln [A]_0} \qquad \textit{Wrong !}$$

The *ln* term cannot be distributed to the numerator and the denominator. The quotient of the expression must first be found and then *ln* applied to the result. You must use "inverse-*ln*" properly to solve this type of problem.

A third formula that you will use for *first order* reactions is the expression for half-life. This expression relates the time for a product to decay to half of its original concentration. The derivation of this relationship is not essential to know, but the formula is. The half-life, $t_{1/2}$ is expressed as:

$$t_{1/2} = \frac{0.693}{k}$$

where k is the rate constant for the *first order* reaction. This equation is found in your textbook in Section 13.5. The solution for either $t_{1/2}$ or k is found by cross multiplication. Examples 13.5 and 13.6 demonstrate the use of this equation. Problems 39–42 involve the use of the half-life equation.

The *second order* reaction relationship of concentration and time is best expressed as:

$$\frac{1}{[A]_t} = kt + \frac{1}{[A]_0}$$

where rate = $k[A]^2$. This equation appears in Section 13.6 of your textbook. It is derived using calculus. You are not required to perform the derivation for these problems, but you must remember the final form of the equation. Note that it is quite different from the *first order* equation.

If you are asked to determine the concentration of A at some time, t, simply insert the data and solve. Remember to use the [1/x] after evaluating the right hand side of the equation.

Other types of problems will involve determining at what time t the concentration of A will decrease from the original value to some lower level. If you are given $[A]_t$, $[A]_o$ and k, the equation can be rearranged to find t as:

$$t = \frac{1}{k} \times \left(\frac{1}{[A]_t} - \frac{1}{[A]_0} \right)$$

Pitfalls in these calculations are the assumption that the concentration data in the brackets can be written as:

$$\frac{1}{[A]_t - [A]_0}$$ *Wrong!*

You must start solving the equation from within the brackets. Use the [1/x] key to calculate the reciprocal of each concentration and then subtract them.

The half-life equation for a *second order* reaction is expressed as:

$$t_{1/2} = \frac{1}{k[A]_0}$$

and is also found in Section 13.6 of your textbook.

Note that the *second order* half-life equation is dependent on the rate constant, while the *first order* equation is independent of k. This information can be most helpful when you are asked to determine the order of a reaction when given only [A] and t data. Example 13.7 and Problems 43, 45, 49, and 50 deal with second order reactions.

9.3 Reaction Rates and Temperatures

One of the larger equations that you will deal with in General Chemistry relates rate constants, temperature, and activation energy. It is the Arrhenius equation and is expressed as:

$$\ln\left(\frac{k_2}{k_1}\right) = \frac{E_a}{R}\left(\frac{1}{T_1} - \frac{1}{T_2}\right)$$

and is found in Section 13.8 of your textbook. A problem using this equation to solve for the *k* value of a given reaction at a higher temperature is demonstrated in Example 13.9 of your text. Problems 55–58 reinforce this concept.

There are several important points to keep in mind when you are working problems with these equations.

- Express the temperature in K.
- Use the gas constant 8.314 J $K^{-}1mol^{-1}$.
- Use e^x to help solve for the *k* value in the *ln* expression (see Section 9.2 of this booklet).
- Take the reciprocals of the temperature before you subtract them.

Section 10 Skills for Chapter 14: Chemical Equilibrium

Goals: to calculate concentrations of products and reactants at equilibrium, and equilibrium constants

Skills: cross multiplication, powers, and roots

10.1 Calculations with Equilibrium Expressions

The equilibrium expression for a hypothetical chemical equation $A + B \rightarrow 2C$ is written as:

$$K_c = \frac{[C]^2}{[A][B]}$$

If you are given the equilibrium concentrations of A, B, and C, the equilibrium constant can be determined by entering these values into the equation. For instance, if [A] = 0.10 M, [B] = 0.20 M, and [C] = 0.50 M, we can write:

$$K_c = \frac{[0.50]^2}{[0.10][0.20]}$$

and

$$K_c = 13$$

Examples 14.1 and 14.2 show the calculation of K_c.

A variation to these problems involves the determination of [C] if [A], [B], and K_c are known. If K_c = 13 with [A] = 0.50 M and [B] = 0.25, let's arrange the above equation to solve for [C].

$$K_c = \frac{[C]^2}{[0.50][0.25]}$$

Next, each side of the equation is multiplied by the denominator, [0.50][0.25]. This act leaves only the term $[C]^2$ on the right hand side of the equation.

$$13 \times ([0.50][0.25]) = [C]^2$$

By rearranging and multiplying the numbers we generate,

$$\sqrt{[C]^2} = \sqrt{1.6}$$

$$[C] = 1.3$$

The equilibrium concentration of C is 1.3 M.

There may be times in solving the equilibrium problems where the determination of a concentration involves taking calculations other than the square root. For instance, let's look at the reaction from p.582 of your text:

$$N_2O(g) + 3/2\ O_2(g) \leftrightarrow 2\ NO_2(g)$$

The equilibrium expression for this reaction is:

$$K_c = \frac{[NO_2]^2}{[N_2O][O_2]^{3/2}}$$

To solve for K_c we must square the NO_2 concentration and then divide by the N_2O concentration and the O_2 concentration raised to the $^3/_2$ power. You will need to use the powers function of y^x on your calculator (see Section 1 of this toolkit).

If, instead, you wish to solve for the O_2 concentration using a given K_c value and known equilibrium concentration of NO_2 and N_2O, the equation must be rearranged.

$$[O_2]^{3/2} = \frac{[NO_2]^2}{[N_2O]K_c}$$

If K_c = 7.9, $[NO_2]$ = 0.30 M, and $[N_2O]$ = 0.050, what is $[O_2]$? Plugging in the numbers, we can write:

$$[O_2]^{3/2} = \frac{[0.30]^2}{[0.050]\ 7.9}$$

$$[O_2]^{3/2} = 0.23$$

$$[O_2] = \sqrt[3/2]{0.23} = (0.23)^{2/3} = 0.23^{0.667}$$

$$[O_2] = 0.37\text{ M} \qquad \text{using the } y^x \text{ function}$$

In order to work any of these types of problems, it is most important that you be able to write the correct equilibrium constant expression for the reaction. If you are in doubt as how to construct the expression, look back at Section 14.2 of your textbook. For more practice with the determination of equilibrium constants, see Problems 19–42.

10.2 Calculation of Equilibrium Concentrations

A second use of the equilibrium expression involves the determination of the concentrations of the products and the reactants at equilibrium. If the stoichiometry of the reaction is known (allowing the equilibrium expression to be written) and K_c is known, then the equilibrium concentrations may be determined.

However, the solution to this problem is not always straightforward. In most cases, we only know the *initial* concentrations of the products and the reactants; we don't know the *equilibrium* concentrations. To calculate these equilibrium concentrations we must figure out how much the concentrations of products and reactants change, based upon the relationship defined by the equilibrium expression.

To help explain how such a solution to this problem is found, we will introduce the "equilibrium checkbook". This teaching tool will allow us to quickly determine the concentrations of all equilibrium products and reactants. The checkbook concept is patterned very much like the checkbook that you are maintaining with a bank in your town. Just as keeping a good record of all deposits and drafts to the account enables you to balance the account, the equilibrium concentration of products and reactants can be readily determined with this system of record keeping.

To see how the checkbook works, let's return to the "alphabet" reaction of Section 10.1 of this toolkit: A + B $\rightarrow$ 2 C. This balanced equation shows that 2 moles of product C are produced when 1 mole of A reacts with 1 mole of B. The equilibrium expression is then written as

$$K_c = \frac{[C]^2}{[A][B]}$$

For the purpose of this example, let's say that $K_c = 0.0100$. The initial concentrations of both A and B are 0.500 M. The initial concentration of C is zero.

In the absence of any initial C, we can easily see that the reaction will proceed toward the right; that is, some C will be produced. However, there are times when you will be asked to perform calculations such as these when the concentration of the "products" (those written on the right-hand side of the equation) are not zero. To determine which direction a reaction will take (e.g., to the right or to the left), you should calculate the reaction quotient, Q (this is discussed in Section 14.3 and Example 14.6 of your text). The formula seems identical to the equilibrium expression, but note that the (initial concentrations) of the products and reactants are used rather than the (equilibrium concentrations).

$$Q = \frac{(C)^2}{(A)(B)}$$

By comparing Q to K_c, the direction of the reaction can be determined.

Problems 49–52 deal with the determination of Q. Problems 53–58 involve the determination of K in a reaction.

In the sample problem for demonstrating the equilibrium checkbook, we have the task of finding [C] from the equilibrium equation. The setup of the checkbook is shown below:

Species	A	B	2 C
initial			
Δ			
equilibrium			

In the top row (labeled species), the reactants and products of the chemical equation are written. The stoichiometry of the equation is also entered in the top row of each column. Note that the double line separates the reactants and products and serves in place of the ↔ symbol. The other rows are labeled: initial, Δ, and equilibrium. These entries are related to the change in the concentration of each reactant and product during the course of the reaction.

With the given information, we can set up the entries in the equilibrium checkbook. Our initial concentrations are entered in the initial row under the proper column of each reactant and product.

Species	A	B	2 C
initial	0.500 M	0.500 M	0 M
Δ			
equilibrium			

The next stage is to determine the changes (withdrawals and deposits) to each account. From the stoichiometry of the reaction, every 1 mol of A and 1 mol of B that react, produce 2 mols of C. With this reasoning, if x mol of A and x mol of B react, then $2x$ mols of C are produced. Since A and B are reactants, their concentrations should be depleted as the reaction proceeds. This "withdrawal" can be entered into the checkbook in the Δ row as $-x$. Similarly, the "deposits" of $+2x$ to the C column can be entered, as shown below:

Species	A	B	2 C
initial	0.500 M	0.500 M	0 M
Δ	$-x$	$-x$	$+2x$
equilibrium			

The last step is to add the activity in each account. This "bottom line" is the equilibrium concentration as a function of initial concentrations and x. By adding the initial and Δ entries together for each product and reactant we find:

Species	A	B	2 C
initial	0.500 M	0.500 M	0 M
Δ	$-x$	$-x$	$+2x$
equilibrium	0.500 M $-$ x	0.500 M $-$ x	$2x$

The entries in the bottom row are the equilibrium concentrations as a function of the initial concentration and x. Since we want to know the equilibrium concentrations as a single number rather than in terms of initial concentrations and x, we must solve for the value of x. Once we know x, we can directly express [A], [B], and [C].

The solution lies in the use of the equilibrium expression. By substituting the data into the equilibrium expression, we can write:

$$0.0100 = \frac{[2x]^2}{[0.500 - x][0.500 - x]}$$

Note that we have one equation and one unknown. To solve for x, we must employ our algebraic skills. For this problem, the solution lies in noting that the equation can be simplified to:

$$0.0100 = \frac{[2x]^2}{[0.500 - x]^2}$$

If we now take the square root of both sides, the x^2 term can be reduced to x, thereby making the equation much easier to solve.

$$\sqrt{0.0100} = \sqrt{\frac{[2x]^2}{[0.500 - x]^2}}$$

which simplifies to

$$0.100 = \frac{[2x]}{[0.500 - x]}$$

By cross multiplication and rearrangement, the following equations show the solution for x:

$$0.100 \times [0.500 - x] = [2x]$$

$$0.0500 - 0.100x = 2x$$

$$0.0500 = 2.100x$$

$$x = \frac{0.0500}{2.100}$$

$$x = 0.024$$

Having found x, the equilibrium values of A, B, and C can be calculated.

$$[A] = (0.500 - 0.024) = 0.476 \text{ mol/L remaining of A}$$

$$[B] = (0.500 - 0.024) = 0.476 \text{ mol/L remaining of B}$$

$$[C] = (2 \times 0.024) = 0.048 \text{ mol/L produced of C}$$

These values are the equilibrium concentrations of the reactants and products. (Note that the setup of this problem can be directly applied to Example 14.12.)

The use of the checkbook will allow you to set up and solve every equilibrium problem that is discussed in your text. Not all problems will be solved by the same exact algebraic methods shown above, but the solutions are similar. The solution to Example 14.13 is a bit different. Since the concentrations of the reactants are not the same, the solution is not obtained by taking the square root of both sides of the equation. Instead, the quadratic formula is used to find the value of x. Additional equilibrium problems are found in Problems 59–68.

The checkbook method can be a powerful tool in calculating equilibrium concentrations. This method can be applied to problems in Chapters 14–16 of your text. The correct use of this equation requires that you must have a balanced equation and an equilibrium expression.

Section

11 Skills For Chapter 15: Acid-Base Equilibria

Goals: to calculate pH values, K_a values, equilibrium concentrations of products and reactants of acid-base reactions, buffers, and the common ion effect.

Skills: multiplication, division, logarithms, powers, roots, and quadratic equations

In this section on aqueous equilibrium, you will be studying methods to calculate the equilibrium concentrations of products and reactants of a chemical reaction. Of the many forms of aqueous equilibrium that exist, acid-base equilibria will be the first in your studies. They are presented first because they can be easily studied in your laboratory section. The basic concepts that you learn in this chapter can be directly applied to the many forms of equilibria that you will encounter in upper level chemistry courses.

11.1 pH Calculations

The value of H_3O^+ in aqueous solution can vary significantly. As your text mentions, the range of these values is influenced by the dissociation of water. Water is capable of producing $[H_3O^+]$ according to:

$$2\ H_2O(aq) \leftrightarrow H_3O^+_{(aq)} + OH^-(aq) \qquad K_w = 1.0 \times 10^{-14}$$

A solution consisting only of water will have a H_3O^+ concentration of 1×10^{-7} M. If a strong acid such as HCl is added to the water, the $[H_3O^+]$ will increase significantly.

Instead of always trying to remember the hydrogen ion concentration of an acid solution as a value expressed to a power, chemists have simplified the values through the concept of pH. By definition:

$$pH = -\log[H_3O^+]$$

If you know the H_3O^+ equilibrium concentration, the pH may be obtained by entering the value of $[H_3O^+]$ into your calculator, pressing the [*log*] key, and then the [+/–] to make the value positive. For example, if the $[H_3O^+]$ is found to be 3.0×10^{-5} M, then the pH is calculated as:

$$pH = -\log [3.0 \times 10^{-5}]$$
$$pH = -(-4.52)$$
$$pH = 4.52$$

Please note that there are two types of logarithmic functions on your calculator, [*log*] and [*ln*]. The pH definition is based on *log* and not *ln*.

We can check the pH calculations with a given $[H_3O^+]$ by using the log relationship of powers to pH.

$$[H_3O^+] = 1.0 \times 10^{-3}$$
$$pH = -\log [1.0 \times 10^{-3}]$$
$$pH = -(-3.00)$$
$$pH = 3.00$$

By similar reasoning, if $[H_3O^+] = 1.0 \times 10^{-5}$, then the pH = 5.00. For the first example worked in this section where $[H_3O^+]$ = 3.0×10^{-5} M, we can predict that the pH should lie between 4 and 5. This second check using estimation can be of aid in confirming the answer from our pocket calculator. A similar example is presented in Problem 39 of your textbook.

Another calculation that will be used in these studies will be the determination of the $[H_3O^+]$ if we are given the pH. This calculation can be performed by using the concept of anti-logs. From our earlier definition,

$$pH = -\log[H_3O^+]$$

$$-pH = \log[H_3O^+] \qquad \text{(multiplying both sides by -1)}$$

$$10^{-pH} = 10^{\log[H_3O^+]} \qquad \text{(using the anti - log relationship)}$$

$$10^{-pH} = [H_3O^+]$$

$$[H_3O^+] = 10^{-pH}$$

To perform these calculations, use the [y^x] or the [10^x] as you did in the previous chapter. If the pH is 5.5 then,

$$[H_3O^+] = 10^{-5.5}$$
$$[H_3O^+] = 3.2 \times 10^{-6}\text{ M}$$

Examine Problem 37 and compare it to the above example.

Note that Problems 40 and 41 as well as other Chapter Problems involve the concept of pOH. The relationship between H_3O^+ and OH^- is discussed in Section 15.3 of your textbook.

You will need to determine whether or not the acid (or base) involved in the reaction is a strong or weak species. As mentioned in your text, the [H_3O^+] for strong acids is usually the initial concentration of the reactants (*i.e.* HCl) because of complete dissociation, whereas the [H_3O^+] for weak acids is determined from the equilibrium expression.

11.2 Equilibrium Calculations

When dealing with weak acids or bases, it is necessary to use the equilibrium expression to calculate the [H_3O^+]. To determine this value, we will return to the concept of an "equilibrium checkbook", outlined in Section 10.2 of this toolkit. This teaching tool allows us to quickly determine the equilibrium concentration of products, [H_3O^+], and reactants.

To start with the method, let's consider how the pH of a 0.10 M solution of acetic acid is determined. The balanced stoichiometric equation is listed below:

$$HC_2H_3O_2(aq) + H_2O(l) \leftrightarrow H_3O^+(aq) + C_2H_3O_2^-(aq)$$

to make our writing a bit more simple, let $OAc^- = C_2H_3O_2^-$.

$$HOAc(aq) + H_2O_{(l)} \leftrightarrow H_3O^+(aq) + OAc^-(aq)$$

The equilibrium expression is written as,

$$K_a = \frac{[H_3O^+][OAc^-]}{[HOAc]}$$

We have the task of finding $[H_3O^+]$ from this equation. All we know is that the initial concentration of HOAc, (HOAc) = 0.10 M and the $K_a = 1.8 \times 10^{-5}$ (from Table 15.2 of your textbook). This information is enough to set up the equilibrium checkbook. We see from the acid dissociation equation that the H_3O^+ and OAc^- must come from the dissociation of HOAc. (Please note that here the assumption is made that the contribution of H_3O^+ from the auto-dissociation of water is negligible as compared to that of the acid.) Any products that are formed, H_3O^+ and OAc^-, must occur from a "withdrawal" from our chemical account that holds the initial "deposit" of the weak acid, HOAc. Also, note that the use of [ ] denotes the equilibrium concentration, while ( ) is used for the initial concentration.

The setup of the checkbook is shown below:

Species	HOAc	H_3O^+	OAc^-
initial	0.10 M	0 M	0 M
Δ			
equilibrium			

The next stage is to determine the changes (withdrawals and deposits) to each account. From the stoichiometry, for each mole of HOAc that undergoes dissociation (withdrawal), one mole of H_3O^+ and one mole of OAc^- is generated (deposits). If x moles of H_3O^+ and x moles of OAc^- are produced, they must have originated from x moles of HOAc (based upon the stoichiometry of the problem). Now that these relationships have been determined, the second row (change) of the checkbook may be entered.

Species	HOAc	H_3O^+	OAc^-
initial	0.10 M	0 M	0 M
Δ	$-x$	$+x$	$+x$
equilibrium			

The last step is to add up the activity in each account. This "bottom line" is the equilibrium concentration as a function of initial concentrations and x.

Species	HOAc	H_3O^+	OAc^-
initial	0.10 M	0 M	0 M
Δ	$-x$	$+x$	$+x$
equilibrium	0.10 M $- x$	x	x

To solve for $[H_3O^+]$, we substitute each of the equilibrium concentrations into the equilibrium expression.

$$K_a = \frac{[x][x]}{[0.10 - x]}$$

Please note that the contribution of $[H_3O^+]$ from the auto-dissociation of H_2O is negligible when dealing with weak acid solutions. Here $[H_3O^+] = [OAc^-]$ because they both came from the dissociation of HOAc. Therefore, the numerator is expressed as x^2. Also, since $K_a = 1.8 \times 10^{-5}$, the equation reduces to:

$$1.8 \times 10^{-5} = \frac{[x]^2}{[0.10\text{ M} - x]}$$

Now, we have one equation with one unknown. There are two methods for the determination of x. The first involves the use of the quadratic equation (found in Appendix A of your textbook). By cross-multiplying and grouping like terms we can write:

$$x^2 + (1.8 \times 10^{-5}x) - (1.8 \times 10^{-6}) = 0$$

Using the quadratic formula, $a = 1$, $b = 1.8 \times 10^{-5}$, and $c = -1.8 \times 10^{-6}$. By plugging these values into the equation, we find x to be 1.3×10^{-3} or -1.35×10^{-3}. Since we cannot have negative values for concentrations, we accept only the first answer.

Let's take a moment to see what we have found: $x = [H_3O^+]$ and $[OAc^-]$, $[HOAc] = 0.10 - (1.3 \times 10^{-3}) = 0.0987$ M (best expressed as 0.10 M using the correct number of significant figures). Note that this change, x, is only 1.3% of the original value. The calculation, which is referred to as the % dissociation is as follows:

$$\% \text{ dissociated } = \frac{[x]}{(\text{HOAc})} \times 100\%$$

Because the change of x in [HOAc] is not significant, we may solve this dissociation equation in a more direct fashion by eliminating x from the denominator.

$$1.8 \times 10^{-5} = \frac{[x]^2}{[0.10 - x]}$$

$$x^2 = 1.8 \times 10^{-6}$$

$$x = 1.3 \times 10^{-3}$$

Care must be taken in the removal of x from the denominator. This removal (which is an approximation to the solution) can only be done if the change in reactant concentration is found to be insignificant. For your studies in acid base equilibrium, this level of insignificance is said to occur when $x < 5\%$ of the original concentration. If you find $x <$ 5% of the original reactant concentration you may use the shortcut. If $x > 5\%$ you must use the quadratic equation to correctly solve for x.

(Of course, you may use the quadratic equation in all of these problems to find x.) However, since the K_a values of the acids you use in your calculations are very small, and you will rarely work with acids much less than 0.001 M in these calculations, you will probably be able to use the shortcut for most of your calculations.

From the definition of equilibrium concentration of the products and reactants, it can be determined that $[H_3O^+]$ and $[OAc^-]$ = 1.3×10^{-3} M and [HOAc] = 0.10 M. The pH of the equation is $-\log[x]$ and is calculated as 2.89.

Example 15.6 is very similar to this problem, where (HOAc) = 1.0 M. Also see Problems 47, 49–51, 53, 55, and 56.

The determination of the pH of solutions of weak bases is shown in Example 15.8, 15.9, and Problems 48, 52 and 54. In some cases the K_b value is not readily available from a set of tables. When this situation is encountered, K_b may be calculated from K_a (from the relationship, $K_w = K_a \times K_b$).

The calculation of the pH of polyprotic acids is demonstrated in Examples 15.11 and 15.12. In the case where $K_{a1} >> K_{a2}$, the acidity of the acid is determined from only K_{a1} (that is, only one equilibrium checkbook is needed). Problems 57–62 also deal with polyprotic acids.

Other applications of K_a and K_b calculations involve the salts of acids and bases (discussed in 15.6 of your text). Examine Examples 15.14 and 15.15 to explore how the checkbook can be applied to these reactions. More practice is found in Problems 63–70.

11.3 Common Ion Effect

In the previous section we studied the equilibrium reaction where acetic acid (HOAc), a weak acid, undergoes dissociation to form:

$$HOAc(aq) + H_2O(l) \leftrightarrow H_3O^+(aq) + OAc^-(aq)$$

If we know the initial concentration of HOAc and the K_a value, we can determine the equilibrium concentrations of H_3O^+ and OAc^-. So far, these problems have only dealt with adding HOAc to pure water (the initial concentrations of the products have been 0). Consider how the reaction would be affected if both HOAc and OAc^-, in the form of sodium acetate (NaOAc), were added to the solution. NaOAc is a strong electrolyte and undergoes complete dissociation. Because,

$$K_a = \frac{[H_3O^+][OAc^-]}{[HOAc]}$$

where $K_a = 1.8 \times 10^{-5}$, the amount of H_3O^+ that is produced is limited by the presence of OAc^- in the reaction vessel. If the initial amount of HOAc is 0.30 M and the initial amount of NaOAc is 0.10 M, we can set up the checkbook to find the $[H_3O^+]$.

Species	HOAc	H_3O^+	OAc^-
initial	0.30 M	0 M	0.10 M
Δ			
equilibrium			

The next step in solving the problem is to write in the activity in this account. From the stoichiometry of this reaction, for every mole of HOAc that undergoes dissociation, a mole of H_3O^+ and a mole of OAc^- is produced. This activity is described as:

Species	HOAc	H_3O^+	OAc^-
initial	0.30 M	0 M	0.10 M
Δ	$-x$	$+x$	$+x$
equilibrium			

Balancing the checkbook yields:

Species	HOAc	H_3O^+	OAc^-
initial	0.30 M	0 M	0.10 M
Δ	$-x$	$+x$	$+x$
equilibrium	0.30 M - x	x	0.10 M + x

Substituting the equilibrium concentrations from the checkbook and the value of K_a allows us to write the expression:

$$1.8 \times 10^{-5} = \frac{[x][0.10 + x]}{[0.30 - x]}$$

To find the $[H_3O^+]$, the equation must be solved for x. If we began to multiply and group like terms, we would have to use the quadratic formula to reach the solution. However, it is possible to make certain assumptions for this reaction that will permit a good approximation of x to be found. In the above equation note that x is in the numerator and denominator. If the change in the equilibrium concentration from adding and subtracting x is insignificant, the expression reduces to:

$$1.8 \times 10^{-5} = \frac{[x](0.10)}{(0.30)}$$

This shortcut can only be used if $x < 5\%$ of the initial reactant concentration. (See Section 11.2 of this toolkit for an explanation.) If we rearrange to solve for x, we can write:

$$\left(1.8 \times 10^{-5}\right)\left(0.30\right) = x\left(0.10\right)$$

$$x = \frac{(1.8 \times 10^{-5})(0.30)}{(0.10)}$$

$$x = 5.4 \times 10^{-5}$$

The value x is the $[H_3O^+]$. Note that this value of x does not alter significantly the concentration of HOAc and OAc^-. The pH of this solution is

$$pH = -\log(5.4 \times 10^{-5})$$

$$pH = 4.27$$

Note that this problem is very similar to Example 15.16 of your text.

To use the approximation in the solution of x, it is most important that the value x not be > 5% of the initial concentration. (This calculation is done in the preceding section of this booklet.) If x > 5%, then the quadratic equation must be used to correctly solve the equation.

As you work with these problems, you will begin to understand when the approximation is valid to use. If your K_a is very small ($< 1 \times 10^{-4}$), and the initial concentrations of reactants and products are large (0.10 to 1 M), you will probably use the approximation. Make it a habit to confirm that $x < 5\%$ of the initial concentrations.

Other examples of the effect of common ions on weak acids can be found in Problems 63–66 of your textbook.

11.4 Buffers

In the above example, the $[H_3O^+]$ produced from the weak acid dissociation is shown to be controlled by the initial concentrations of reactant and products and the equilibrium constant. In your text, an equation is described that relates the pH (equilibrium H_3O^+) to the concentration of the conjugate acid-base pairs. It is called the Henderson-Hasselbalch equation and is written as:

$$pH = pK_a + \log \frac{[\text{conjugate base}]}{[\text{weak acid}]}$$

and is found in Section 15.8 of your textbook.

More commonly, this expression is known as the buffer equation. By adjusting the (conjugate base)/(weak acid) ratio of a weak acid-conjugate base pair, the pH can be controlled. Using this equation, go back to the previous problem and solve for the pH of the mixture. From the given data $(OAc^-) = 0.10$ M and $(HOAc) = 0.30$ M. To find the pK_a value we take the negative log of K_a.

$$pK_a = -\log K_a$$
$$pK_a = -\log(1.8 \times 10^{-5})$$
$$pK_a = -(-4.74)$$

$$pK_a = 4.74$$

To solve for the pH, enter the data into the buffer equation.

$$pH = 4.74 + \log\frac{[0.10]}{[0.30]}$$

$$pH = 4.74 + (-0.48)$$

$$pH = 4.26$$

Problem 81 in your textbook is similar to this example. A checkbook approach is used to find $[H_3O^+]$. Other work with buffers is found in Problems 81–88.

An important use of this equation is to calculate the ratio of (conjugate base)/(weak acid) required to make a buffer of a specific pH. Let's say you wished to make a buffer solution where the pH = 4.50. If you were using the HOAc and NaOAc conjugate pair to make the buffer solution, in what ratio would you mix them? We can rearrange the buffer equation to solve for the ratio.

$$pH = pK_a + \log\frac{[\text{conjugate base}]}{[\text{weak acid}]}$$

$$\log\frac{[\text{conjugate base}]}{[\text{weak acid}]} = pH - pK_a$$

$$\frac{[\text{conjugate base}]}{[\text{weak acid}]} = 10^{(pH - pK_a)}$$

Entering the data into this equation, we can write:

$$\frac{[\text{conjugate base}]}{[\text{weak acid}]} = 10^{(4.50-4.74)}$$

$$\frac{[\text{conjugate base}]}{[\text{weak acid}]} = 10^{-0.24}$$

$$\frac{[\text{conjugate base}]}{[\text{weak acid}]} = 0.58$$

The conjugate base to weak acid ratio should be 0.58 to achieve a pH of 4.50. If you wished to use a (HOAc) of 0.10 M, the (OAc^-) would need to be 0.058 M. Use this approach to solve Example 15.18 and Problems 83 and 84.

NOTE: Remember that log (1) = 0. If the ratio of the conjugate base to weak acid is 1, the pK_a of the weak acid-conjugate base pair is the pH.

In Chapter 15 of your textbook, you are introduced to the calculation of the pH of a buffered solution when a strong acid or base is added to the buffer solution. To solve a problem like this, we will first have to use the checkbook and then the buffer equation. For this example, let's look at the effect of adding 0.02 M HCl to a buffered solution containing 0.10 M HOAc and 0.10 M NaOAc. (Since the (conjugate base)/(weak acid) = 1, the pH before the HCl is added to the solution is the pK_a value (of 4.74.) The checkbook for the balanced equation is started as:

Species	HOAc	H_3O^+	OAc^-
initial	0.10 M	0.02 M	0.10 M
Δ			
equilibrium			

Note that the [H_3O^+] is not 0 M here, but the concentration of the strong acid that is added to the buffer solution. This addition should force the reaction to the left; that is, excess H_3O^+ will react with OAc^- to produce HOAc. Filling out the change columns yields:

Species	HOAc	H_3O^+	OAc^-
initial	0.10 M	0.02 M	0.10 M
Δ	$+x$	$-x$	$-x$
equilibrium			

Completing the checkbook, we can write,

Species	HOAc	H_3O^+	OAc^-
initial	0.10 M	0.02 M	0.10 M
Δ	$+x$	$-x$	$-x$
equilibrium	$0.10\text{ M} + x$	$0.02\text{ M} - x$	$0.10\text{ M} - x$

A driving force in this aqueous buffer chemistry problem is that the strong acid added to the solution reacts completely with the OAc^- to form HOAc. If the amount of strong acid added to the solution is 0.02 mol/L, then the x term in the above equation is 0.02 M. The H_3O^+ concentration will be 0 M after the reaction is over. (Note: an assumption in your text is that H_3O^+ is describing the strong acid concentration. After the reaction is over, the strong acid should be "neutralized" and its concentration 0 M. This does not imply the absence of H_3O^+ ions in solution. The HOAc will provide a small amount, as will the solvent, of water. But for our purpose of buffer calculations, the concentration is 0 M.)

From the final row of the checkbook, it can be determined that [HOAc] = 0.12 M and [OAc^-] = 0.08 M. If we place these numbers into the buffer equation, we can calculate the pH.

$$\text{pH} = 4.74 + \log\frac{[0.08]}{[0.12]}$$

$$\text{pH} = 4.74 + \log(0.67)$$

$$\text{pH} = 4.74 + (-0.18)$$

$$\text{pH} = 4.56$$

Remember: It is a good practice to check the logic of the calculation. If you add a strong acid to a buffer solution, the pH should decrease from the original value. If you add a strong base to a buffer solution, the pH should increase. If the opposite, you have probably reversed the concentrations in the conjugate base/weak acid ratio.

Try Problems 85 and 86 with this approach. They can be solved with the equilibrium checkbook.

This approach to buffers is also useful when titrating a weak acid with a strong base (*e.g.* titrating HOAc with NaOH). At the start of the titration, the only OAc^- present is from the weak acid dissociation. When the strong base is added to the solution the following reaction occurs:

$$HOAc + OH^- \rightarrow OAc^- + H_2O$$

This reaction proceeds to completion; that is, for every 1 mol of OH^- that is introduced into the HOAc solution, 1 mol of HOAc is consumed and 1 mol of OAc^- is produced. Since OH^- is added in significant quantities, the $[OAc^-]$ is primarily the result of the titration. The amount of OAc^- (x) from the weak acid dissociation is negligible when compared to the amount produced by the titration reaction. Hence, the titration produces the buffer solution. The pH of this solution is found using the buffer equation.

The calculation of pH requires us to find the [HOAc] and $[OAc^-]$ resulting from the titration. Consider a 50.0 mL solution of 0.100 M HOAc to which 25.0 mL of a 0.0700 M standardized base solution of NaOH has been added. The original mol amount of HOAc is 0.100 M × 0.0500 L = 0.00500 mol. The mol amount of OH^- that has been added to the solution is 0.0700 M × 0.0250 L = 0.00175 mol. The mol amount of HOAc remaining is 0.00500 - 0.00175 = 0.00325 mol. The mol amount of OAc^- produced is 0.00175 mol.

To use the buffer equation we must express OAc^- and HOAc in terms of concentration. This requires us to express each in terms of molarity. To do this, we must first calculate the total volume of the solution: 50.0 mL of weak acid + 25.0 mL of base = 75.0 mL, or 0.075 L. Next, introduce these data into the buffer equation.

$$pH = pK_a + \log\left(\frac{\dfrac{0.00175 \text{ mol OAc}^-}{0.075 \text{ L}}}{\dfrac{0.00325 \text{ mol of HOAc}}{0.075 \text{ L}}}\right)$$

$$pH = pK_a + \log(0.538)$$

$$pH = 4.74 + (-0.27) \qquad \text{since } pK_a = 4.74$$

$$pH = 4.47$$

Use this approach to solve Example 15.21, Problems 97, 99, and 100.

Section 12 Skills for Chapter 16: Equilibria Involving Slightly Soluble Salts and Complex Ions

Goals: to calculate solubility and K_{sp}, to examine the effects of pH, common ions, and complex ions on solubility

Skills: multiplication, division, logarithms, powers, roots and quadratic equations

12.1 Solubility

The equilibria presented in this chapter are based upon the concept of the solubility of salts and ions in aqueous media. Consider the process of dissolving AgCl in water.

$$AgCl(s) \leftrightarrow Ag^{+}(aq) + Cl^{-}(aq) \qquad K_{sp} = 1.8 \times 10^{-10}$$

Since this is a heterogeneous reaction, the equilibrium expression is written as:

$$K_{sp} = [Ag^{+}][Cl^{-}]$$

The AgCl is a solid and will not enter into the equilibrium calculation or the checkbook.

The K_{sp} expression is useful in determining the molar solubility of a compound. If excess AgCl is placed in water, the K_{sp} expression will define the maximum allowable product of two species. We can see how this works if we set up the checkbook.

Species	Ag^{+}	Cl^{-}
initial	0 M	0 M
Δ		
equilibrium		

Because $AgCl(s)$ is not part of the heterogeneous equilibrium expression, it is not listed here. (Remember, the concentration of a solid cannot change, therefore only the soluble species in this reaction are listed.)

The change row is determined from the stoichiometry of the equation. Here 1 mole of $AgCl(s)$ yields 1 mole of $Ag^+(aq)$ and 1 mole of $Cl^-(aq)$. If x mol/L of AgCl dissolve, then the checkbook becomes:

Species	Ag^+	Cl^-
initial	0 M	0 M
Δ	$+x$	$+x$
equilibrium		

The equilibrium concentrations are calculated as:

Species	Ag^+	Cl^-
initial	0 M	0 M
Δ	$+x$	$+x$
equilibrium	$+x$	$+x$

If we substitute these values for the equilibrium concentrations into the equilibrium equation, we obtain:

$$K_{sp} = [x]\,[x] \qquad \text{or}$$

$$K_{sp} = [x]^2$$

$$x = \sqrt{K_{sp}}$$

Solving for x will tell us the molar concentrations of each ion in solution. Plugging in K_{sp} and solving this expression yields:

$$x = \sqrt{1.8 \times 10^{-10}}$$

$$x = 1.3 \times 10^{-5}$$

The molar solubility of AgCl(s) is 1.3×10^{-5} M. Compare this problem to Example 16.3 using $Ag_2SO_4(s)$ and note how the equilibrium checkbook may be applied.

The stoichiometry of the equation plays an important part in determining the molar solubility of a compound. Consider the salt, CaF_2. The balance dissociation equation and equilibrium constant are listed below:

$$CaF_2(s) \leftrightarrow Ca^{2+}(aq) + 2\,F^-(aq) \qquad K_{sp} = [Ca^{2+}][F^-]^2$$

The completed checkbook is listed below for excess $CaF_2(s)$ dissolved in water.

Species	Ca^{2+}	$2\,F^-$
initial	0 M	0 M
Δ	$+x$	$+2x$
equilibrium	$+x$	$+2x$

Note that 2 moles of F^- are produced for every 1 mole of $CaF_2(s)$ that dissolves. If we enter these values into the equilibrium expression we obtain:

$$K_{sp} = [x][2x]^2$$

Because $K_{sp} = 5.3 \times 10^{-9}$ (from Table 16.1 and Appendix C of your textbook).

$$5.3\times10^{-9} = [x][2x]^2$$

To determine the molar solubility, we must rearrange the expression in terms of x. Note that the $2x$ term is squared. We must deal with this term first before we rearrange.

$$5.3\times10^{-9} = \left[x\right]\left[4x^2\right] \qquad \text{(inserting terms)}$$

$$5.3 \times 10^{-9} = \left[4x^3\right] \qquad \text{(combining terms)}$$

$$\left[x^3\right] = \frac{5.3 \times 10^{-9}}{4} \qquad \text{(dividing by 4)}$$

$$x = \sqrt[3]{1.3 \times 10^{-9}} \qquad \text{(taking cube root)}$$

$$x = 1.1 \times 10^{-3} \qquad \text{(the molar solubility)}$$

See Problems 23–36 for more work concerning solubility.

If you have difficulty taking the cube root of this expression or your calculator does not have a cube root function, see Section 1 of this toolkit. Also, while setting up the problem, make sure you have raised all concentrations to the appropriate power before you begin to rearrange the equation and solve for x.

12.2 Effect of Common Ions and pH

Your text also discusses the effect of common ions and pH of the solubility of slightly soluble salts. These effects can be calculated through the use of the equilibrium checkbook, as outlined in Section 10.2 of this toolkit. See Examples 16.5–16.10 and Problems 37–46 for work with common ions and Example 16.11 and Problems 63–68 for pH. In each case, start with the balanced chemical equation, determine how the effect will shift the equilibrium, and solve for x in the checkbook.

12.3 Complex Ions

The solubility of salts can be greatly enhanced by the formation of complex ions. As discussed in Section 16.6 of your text, the reactions have extremely large equilibrium constants, K_f. This large value of K_f, as opposed to K_{sp}, causes us to introduce an assumption into the equilibrium checkbook.

This assumption allows us to start the checkbook with the maximum amount of product that can be formed. For instance, in the following reaction

$$Ag^{+}(aq) + 2\,NH_3(aq) \leftrightarrow [Ag(NH_3)_2]^{+}(aq) \qquad K_f = 1.6 \times 10^{7}$$

if 1 mol of Ag^+ and 2 mol of NH_3 were combined in a beaker, 1 mol of $[Ag(NH_3)_2]^+$ would be quantitatively formed because of the large K_f. We assume that the reaction goes to completion. Any free Ag^+ that remained in solution would essentially be a result of the reverse reaction of the equilibrium process.

To examine this principle, consider a solution that initially contained 0.50 M $AgNO_3$ and 2.0 M NH_3. What is the concentration of Ag^+ in solution after equilibrium has been reached?

The solution to this problem begins with the verification of a balanced chemical equation and the writing out of the equilibrium expression. Using the balanced equation above, the K_f is written as,

$$K_f = \frac{[Ag(NH_3)_2]^+}{[Ag^+][NH_3]^2}$$

The checkbook is constructed and the initial entries are made next.

Species	Ag^+	2 NH_3	$[Ag(NH_3)_2]^+$
initial	0	1.0 M	0.50 M
Δ			
equilibrium			

Just as with the stoichiometric example, a solution containing 0.5M Ag^+ in excess NH_3 will form 0.5 M of the product. Hence, the "initial" amount of the Ag^+ is 0 M and the product is 0.50 M. The "initial" amount of NH_3 is the concentration that was not used (*e.g.* 2 M of original NH_3 – 1 M NH_3 that reacted with the Ag^+ to form the product = 1.0 M remaining.)

The setup of the checkbook allows the direction the reaction proceeds in to establish equilibrium to be clearly seen; the reverse reaction will reestablish the equilibrium concentration of Ag^+. This step is now entered into the equilibrium checkbook.

Species	Ag^+	$2\ NH_3$	$[Ag(NH_3)_2]^+$
initial	0	1.0M	0.50 M
Δ	$+x$	$+2x$	$-x$
equilibrium			

The final operation involves the expression of the equilibrium concentrations.

Species	Ag^+	$2\ NH_3$	$[Ag(NH_3)_2]^+$
initial	0	1.0 M	0.50 M
Δ	$+x$	$+2x$	$-x$
equilibrium	$+x$	$1.0 + 2x$ M	$0.50 - x$ M

By entering these values into the equilibrium expression, we write,

$$1.6 \times 10^7 = \frac{0.50 - x}{x[1.0 + 2x]^2}$$

As with acid-base equilibria problems, the effect of x on the change in concentration of the major species present in the reaction is negligible (as long as x is < 5% of the initial concentration). If x is eliminated from the NH_3 and $[Ag(NH_3)_2]^+$ concentrations, the equation simplifies to

$$1.6 \times 10^7 = \frac{0.50}{x[1.0]^2}$$

Solving this problem via cross multiplication and division yields

$$x = \frac{0.50}{1.6 \times 10^7}$$

$$x = 3.1 \times 10^{-8}$$

Therefore, the equilibrium concentration of Ag^+ is a very small 3.1×10^{-8} M.

Compare this problem to Examples 16.13–16.15 and Problems 77–84 of your text.

Section

13 Skills for Chapter 17: Thermodynamics

Goals: to calculate entropy, enthalpy, and free energy changes and to calculate the effect of free energy on the equilibrium constant

Skills: multiplication, division, logarithms, and powers

13.1 Enthalpy and Entropy Changes

In Chapter 6 of your text, the concept of enthalpy was presented to you. Each product and reactant in a chemical reaction can be assigned an enthalpy value, ΔH°_f (found in Appendix C of your text). The change in enthalpy, ΔH, is calculated by subtracting the sum of the enthalpy values of the reactants from the sum of enthalpy values of the products. The standard entropy change can be calculated in an identical fashion. The S° values for certain species are also listed in Appendix C. Remember, the ° symbol means that all products and reactants involved in the reaction are at standard state (gases at 1 atm pressure, solutions at 1 M, T = 298 K). See Example 17.3 and Problems 41 and 42.

13.2 Gibbs Free Energy

The change in free energy is defined as:

$$\Delta G = \Delta H - T\Delta S$$

This ΔG term can be calculated by two methods. In Appendix C of your text, ΔG°_f values are listed. The change in free energy, ΔG, is calculated by subtracting the sum of ΔG°_f values of the reactants from the sum of ΔG°_f values of the products. The second method would involve using the $\Delta H°_f$ and $\Delta S°$ values for the reaction and the temperature under which the reaction was run. Please note that the units of $\Delta H°_f$ are kJ/mol while $\Delta S°$ values are in J mol^{-1} K^{-1}. You will need to convert these values to either J or kJ to add them together and find $\Delta G°$. For more practice, try Example 17.6 and Problems 39–44.

13.3 Free Energy and Equilibrium

The relationship that is used to relate the $\Delta G°$ and the equilibrium constant is:

$$\Delta G° = -RT \ln K_{eq}$$

Here R is the gas content [in J mol^{-1} K^{-1}], T is the temperature in Kelvin. This equation is described in Section 17.6 of your textbook. You will use this equation to determine the equilibrium constant K_{eq} when the $\Delta G°$ value is given or determined from tables.

To determine the K_{eq} value using this equation, it is necessary to rearrange the equation for K_{eq}, and take the anti-log of both sides of the equation. This is shown below:

$$e^{\left(\frac{-\Delta G°}{RT}\right)} = e^{\left(\ln K_{eq}\right)} \qquad \text{(which reduces to)}$$

$$e^{\left(\frac{-\Delta G°}{RT}\right)} = K_{eq}$$

Remember: Whenever an *ln* term is used in an e^x expression, the two terms cancel each other out.

For our above example, K_{eq} for a reaction can be found where $\Delta G° = -1.20$ kJ at 23°C . Before entering the data into the equation, the values must be listed in the appropriate units.

$$\Delta G° = -12{,}000 \text{ J}$$

$$T = 296\ \text{K}$$
$$R = 8.314\ \text{J mol}^{-1}\ \text{K}^{-1}$$

Now, substituting into the equation:

$$e^{\left(\frac{-(-12,000)}{8.314 \times 296}\right)} = K_{eq}$$

$$e^{(4.88)} = K_{eq} \qquad \text{(using the } e^{x} \text{function)}$$

$$K_{eq} = 130$$

Example 17.9 in your textbook shows another problem of this type of calculation for determining K_{eq}. See Problems 49–60 for more work.

Another type of calculation that you may encounter is the determination of $\Delta G°$ if you are given K_{eq}, and T. For instance, calculate $\Delta G°$ for a reaction which has a K_{eq} of 2.0×10^{3} at 25°C. Substitute this information into the formula and write:

$$\Delta G° = -[8.314\ \text{J mol}^{-1}\ \text{K}^{-1}]\ [298\ \text{K}]\ [\ln 2.0 \times 10^{3}]$$

and simplifying

$$\Delta G° = -[8.314\ \text{J mol}^{-1}\ \text{K}^{-1}]\ [298\ \text{K}]\ [7.6]$$
$$\Delta G° = -1.9 \times 10^{4}\ \text{J}$$

Review Problems 49a and 50a for the calculation of $\Delta G°$. Problems 61–70 involve $\Delta G°$ and K_{eq} as functions of temperature.

Section

14 Skills for Chapter 18: Electrochemistry

Goals: to calculate K values based on cell potentials, to calculate cell potentials using the Nernst equation, and to calculate the amount of metals deposited during electrolysis

Skills: multiplication, division, logarithms

14.1 emf, *K*, and the Nernst Equation

An electrochemical cell contains reactants and products that are undergoing a chemical reaction. As we have learned in the previous chapters, the concentration of products and reactants in a chemical reaction at equilibrium can be expressed through the use of the equilibrium constant and equilibrium expression. In this chapter of your textbook, the authors have shown you new equations that relate the cell potential (E) to the free energy change (ΔG), and the reaction quotient (Q) to ΔG. Through substitution of these two relationships, the Nernst equation was obtained (see Chapter 18.6). The equation is shown below:

$$E_{\text{cell}} = E^{\circ}_{\text{cell}} - \frac{2.303RT}{nF}\log Q$$

Here R is the gas constant (8.314 J K^{-1} mol^{-1}), T is the temperature in K, n is the number of electrons involved in the reaction, *and* F is the Faraday constant of 96,500 J V^{-1} mol^{-1}. You may note that the equation has several constants. We can simplify this equation by reducing (R/F) to one number. To make the expression even simpler, reactions run at room temperature (298 K) allow us to factor in the temperature, T. Also we can include the ln–log conversion term of 2.303. Hence, we may reduce (2.303 RT/F) to a value of 0.0592 V. (Note that the unit of the reduced term is volts.)

$$E_{cell} = E°_{cell} - \frac{0.0592}{n} \log Q \qquad (T = 298\ K)$$

You may be asked to use this expression to calculate E_{cell} for a chemical reaction. For instance, let's say a chemical reaction possessed a Q of 5.0×10^{-10} and an $E°_{cell}$ of 0.25 V. Also, the reaction involved a 1 electron transfer ($n = 1$). Based on this information, E_{cell} can be calculated with the above formula:

$$E_{cell} = E°_{cell} - \frac{0.0592}{1} \log(5.0 \times 10^{-10})$$

$$E_{cell} = 0.25 - 0.0592 \times (-9.30)$$

$$E_{cell} = 0.25 - (-0.55)$$

$$E_{cell} = 0.80\ V$$

A good example of how to use the Nernst Equation can be seen in the determination of cell potential of a "Cu/Zn" battery at 25.0°C. The redox reaction is:

$$Zn^0(s) + Cu^{2+}(aq) \leftrightarrow Zn^{2+}(aq) + Cu^0(s)$$

We can determine that $E°_{cell} = 1.10$ V from the tables of standard reduction potentials and $n = 2$. If we are told that the $(Cu^{2+}) = 0.20$ M and $(Zn^{2+}) = 0.80$ M, the calculation can be started. First, let's determine Q. The expression for Q in this reaction is:

$$Q = \frac{(Zn^{2+})}{(Cu^{2+})}$$

The Q expression will only be concerned with the concentration of those species in solution. The solid Zn and Cu (the electrodes) are not involved in the calculation of Q. Using the solution data we find:

$$Q = \frac{(0.80)}{(0.20)} = 4.0$$

Entering these data into the Nernst equation, we write:

$$E_{\text{cell}} = 1.10 - \frac{0.0592}{2} \log 4$$

$$E_{\text{cell}} = 1.10 - \frac{0.0592}{2} \times (0.602)$$

$$E_{\text{cell}} = 1.10 - (0.0296) \times (0.602)$$

$$E_{\text{cell}} = 1.08 \text{ V}$$

If you are given concentrations of the products and reactants in the redox equations and you know the temperature, you can calculate the cell potential. Of course, you have to be able to write the balanced redox equation. This step is most crucial, since the Q expression and n value are dependent on the stoichiometry of the balanced redox reaction.

Example 18.9 is similar to the example shown above. Problems 57–60 are also similar. Problems 61 and 62 are variations on this equation where you are asked to find the concentrations of products or reactants based upon a given E_{cell} value.

In many situations we may wish to determine the value of the equilibrium constant K rather than Q. This can be accomplished by knowing that at equilibrium $\Delta G = 0$ and therefore, $E_{\text{cell}} = 0$ for a reaction. (Verify this for yourself using $\Delta G = -nFE_{\text{cell}}$.) We can substitute this information into the above equation and find:

$$0 = E^\circ_{\text{cell}} - \frac{0.0592}{n} \log K \quad \text{(substituting values)}$$

$$E^\circ_{\text{cell}} = \frac{0.0592}{n} \log K \quad \text{(rearranging)}$$

$$\log K = \frac{nE^\circ_{cell}}{0.0592} \qquad \text{(solving for } K\text{)}$$

As we have seen earlier, to solve for K we must take the anti–log (10^x of both sides of the equation). Doing so generates the following equation:

$$K = 10^{\left(\frac{n E^\circ_{cell}}{0.0592}\right)}$$

As an example, consider a reaction where $n = 2$ and $E^\circ_{cell} = 1.00$. The value of K is calculated as follows:

$$K = 10^{\left(\frac{2 \times 1.00}{0.0592}\right)}$$

$$K = 10^{(33.8)}$$

$$K = 7 \times 10^{33} \qquad \text{(by using } 10^x\text{)}$$

Problems 53 and 54 involve the use of this equation to find K_{eq} for a reaction. These problems are solved by first balancing the chemical equations (to find the value of n) and calculating E°_{cell}. With these data, K_{eq} can be calculated.

You may also be asked to calculate E°_{cell} if given the K_{eq} value. Enter the data into the formula, (with care to use the *log* button and not *ln*) for the determination of K_{eq}. Also, see if the temperature of the reaction is 298 K (25°C).

14.2 Electrolysis

An important application of electrochemistry is electrolysis, sometimes known as electroplating. Because the chemical reaction used in this process is based on a redox reaction, the monitoring of the moles of electrons passed through the reaction vessel will inform us as to the number of moles of material plated on the electrodes. To determine the number of moles of electrons, the number of amperes (A) is multiplied by the number of seconds of current flow and then divided by F (Faraday's constant).

$$n_{e^-} = \frac{A \times t}{F}$$

After obtaining the n_{e^-} value, the stoichiometry of the reduction half reaction allows us to see the theoretical mole quantity of metal that would be deposited.

Sometimes the electrolysis reactions take several minutes to hours to perform. This time value is greatly dependent on the current used. Therefore, take care in working these problems so that all time values are converted to seconds.

See Examples 18.13, 18.14 and Problems 75–84 for more work with electrolysis.

Section

15 Skills for Chapter 19: Nuclear Chemistry

Goals: to determine the half-life of a radioactive element
Skills: multiplication, division, logarithms, exponential notation

The half-life equation you explored in Chapter 13 of your text can be used further to predict the decay process of radionuclides. This decay process follows first-order kinetics. As your text states in Section 19.3, the equations are:

$$\ln \frac{N_t}{N_0} = -\lambda t \qquad \text{(decay equation)}$$

$$t_{1/2} = \frac{0.693}{\lambda} \qquad \text{(half-life equation)}$$

To illustrate the use of these equations, consider the medical application of radioactive tracers that are used to monitor illnesses in patients. If a certain tracer has a half-life of 30.0 minutes, how much of a 4.00 mg sample of the tracer would still be active after 45.0 minutes?

First, let's determine λ. From the half-life data and the equation:

$$30.0 \text{ min} = \frac{0.693}{\lambda}$$

$$\lambda = 0.0231 \text{ min}^{-1}$$

With this value of λ and N_0 of 4.00 mg, we can use the decay equation to solve for the remaining mass of the tracer at $t = 45.0$ min, $N_{45.0}$:

$$\ln \frac{N_{45.0}}{4.00 \text{ mg}} = -(0.0231 \text{ min}^{-1}) \times (45.0 \text{ min})$$

$$\ln \frac{N_{45.0}}{4.00 \text{ mg}} = -1.0395$$

$$\frac{N_{45.0}}{4.00 \text{ mg}} = e^{-1.0395} \qquad \text{(using } e^x \text{ to remove } \ln \text{ x)}$$

$$\frac{N_{45.0}}{4.00 \text{ mg}} = 0.3536$$

$$N_{45.0} = 1.41 \text{ mg}$$

Example 19.2 and Problems 33 and 34 are similar to this problem. Problems 37 and 38 deal with calculating the half-life of isotopes.

Note that Example 19.4 and Problems 41 and 42 deal with the "dating" of samples. This type of problem can be solved if N_o and $t_{1/2}$ of the element contained in the sample are known and if N_t can be measured. Using the rate law, t can be found.

Section 16

Self Test for Math Skills

Listed below are a series of questions designed to check your math skills. Please take no more than 30 minutes in answering them. You will find the answers and an assessment chart in the back of this booklet.

1. How many significant figures are in the number 0.101?

2. Solve for x in the following equation with care to express the correct number of significant figures.

$$x = 12.011 + 1.00797$$

3. Solve for x in the following equation with care to express the correct number of significant figures.

$$x = \frac{1.057}{10.3}$$

4. Express the number 0.000356 in scientific notation.

5. Determine the mean, $\bar{x}$, for the following data: 25.12, 25.29, 24.95, 25.05 and 25.55.

6. Solve the equation: $x = 2.0 (\sin 35°)$.

7. Solve the equation: $0.405 = \cos(x)$.

8. Determine the value of x in the following equation.

$$(20.3) \times (0.0171) = \left(\frac{x}{60.3}\right) \times (1.35 \times 10^3)$$

9. What percentage of 15.5 is 1.25?

10. Solve the following equation for x.

$$\frac{1}{x} = \frac{4.3}{6.8}$$

11. Solve the following equation for x.

$$\frac{0.57}{0.22} = \frac{1.2}{x^2}$$

12. Calculate the value of x, where $x = (6.23 \times 1.22)^3$.

13. What is the fourth root of 17?

14. Solve the following equation for x.

$$x = 8.3 \times \left(\frac{1}{2^2} - \frac{1}{4^2}\right)$$

15. Determine the value of x in the following equation.

$$2.52 = \sqrt{\frac{1}{x}}$$

16. Calculate the value of x, where $x = \log(0.017)$.

17. Find the value of x, where $x = e^{\ln(12)}$.

18. Determine the value of x, where $1.53 = \log(x)$.

19. Solve for x in the following equation.

$$x = \frac{-1}{0.030} \times \ln\left(\frac{0.20}{0.50}\right)$$

20. Solve for x in the following equation.

$$x = \left(1.00 \times 10^{3}\right) e^{\left(\frac{-0.0220}{1.33}\right)}$$

Section

17 Chemistry and Writing

17.1 The Scientific Notebook

The scientific notebook is the scientist's own record of experiments performed and phenomena observed. Beginning with the first student laboratory report there are special requirements for recording experimental results. The requirements may seem rigid at first, but they are very understandable in the light of the purposes of the notebook.

For the professional scientist the claim to original work is found in the scientific notebook. Millions of dollars in patent rights may depend on the existence of a properly dated and authenticated scientific notebook. Many of the rules that are followed in recording data follow from this important function of the notebook. Nothing is ever erased; and incorrect reading is crossed out and the correct one written beside it. Work is recorded in a bound notebook with pages that cannot be removed or added. Every entry is dated, signed, and countersigned by the scientist in charge of the laboratory. All these rules are designed to produce a record that will constitute proof not only of what experiments were performed, but of the exact date. This is important, because if two scientists make the same discovery, the first one to do so will gain all the legal rights and most of the credit for the work. Obviously, it is more important to have a complete and original record than a perfectly neat one. A few crossed-out readings are not uncommon, and a few blots from spilled chemicals are not unheard of either. These are preferable in the laboratory notebook to a perfect page that has been copied over at a later date and no longer constitutes an authentic original record. **Under no circumstances is data to be recorded on loose paper rather than directly into the notebook!**

Another important function of the notebook is to record the procedure and observations so clearly and completely that the experiment can easily be repeated at a later date. Experiments that cannot be repeated by the same researcher or by other laboratories are soon discredited. For the student in the laboratory complete notes are important as well. If something goes wrong, it should be possible to find the error in procedure from the experiment notes. At times, the numbers in the crossed-out data entries tell an interesting story. Occasionally an interesting and unexpected phenomenon will be observed that merits further study. Always a complete, clear record of what has happened in the laboratory is essential.

In order to be a complete record, each experiment entered into the notebook should include certain features. The scientist's **name** and the **date** should always be entered. The **title of the experiment** being performed is an important element that is often neglected. "Chemistry Lab" is an inadequate substitute for the experiment title, which is usually readily available. Often it is useful to begin by writing the **objective**, or the purpose, of the experiment. Stating the objective clearly helps both the experimenter and the reader of the notebook to understand the experiment. A complete record of experimental **procedure** is essential, either as a step-by-step description or by a complete reference to a standard experimental procedure. If a standard procedure is given, great care must be made to note any deviations from that procedure. A list of **materials and equipment** used can be a great help in organization if it is included as a part of the experimental procedure.

Though the laboratory notebook does not have to be perfectly pristine, it is certainly desirable that it should be as organized as possible. Some time and thought spent in planning before the laboratory period begins will result in a better notebook and a more successful experiment. **The date, title, experimenter's name and objective of the experiment should be entered before the experiment begins.** If the experimental procedure that has been provided does not already give **labeled data tables** for an experiment, it is worth some time and thought to set up such tables before entering the laboratory, rather than waste time during the experiment deciding how to do so. Ample space should be provided not only for the expected data, but also for corrections and notes. Unused space can be crossed out later as necessary, though extra pages are never torn out. Sometimes only the right-hand pages of the notebook are used, leaving the other pages free for later notes or calculations. Individual research laboratories or student laboratories may have standard notebooks or forms in which to write laboratory results. All of them share the basic objective of recording in a useful way the scientist's actions, observations, and thoughts while in the laboratory.

17.2 The Scientific Report

When the scientist makes a formal written report of experiments performed in the laboratory, the report follows a generally accepted outline. Introduction, results and discussion, conclusions follow in order as separate sections and are clearly labeled. Lists of references and even the title are treated in standard ways.

The **title** of a scientific paper is seldom an occasion for creativity. Titles for articles in scientific journals are carefully constructed from words that will be useful key words for information searches by computer. Titles for student laboratory reports are usually indicated in the assignment. As with the laboratory notebook, "Chemistry Laboratory" is unacceptably vague as a laboratory report title. Abbreviations as part of a title should be avoided.

The **Introduction** section should make clear to the reader the purpose and the background of the experiment. The objective of the work that is being discussed should always be clearly stated. It may be appropriate to discuss the basis of the experimental methods that were used as well as the scientific theory on which the work is based. Usually a well-written introduction makes use of written resources in the form of scientific books and papers, which must be listed in the references cited and footnoted with the appropriate reference.

The **Experimental Procedure** section explains in detail exactly how the experiment was conducted. It should be possible to reproduce the experiment using the information found in this section. If standard procedures are used and not explained in detail, a reference should be given. A list of materials and equipment is often a useful component in this section. It includes all chemicals used, including the concentrations of solutions, and all special equipment.

The **Results and Discussion** section includes the data that were obtained in the experiment together with an explanation of the data. Often it is useful to organize the results of the experiment in tables, and sometimes graphs are required as well. All tables and figures should be titled and numbered. All columns in tables and both axes of a graph should be carefully labeled, not omitting units. If calculations have been performed, the equations used should be clearly indicated and enough information about the calculations should be included so that they can be clearly followed. The precision and accuracy of the results should be calculated by standard statistical methods if appropriate to the experiment.

The **Conclusions** section contains the thoughts of the experimenter about the significance of the work performed. Each part of the experiment should be discussed. Numerical results should be evaluated, and the meaning of any statistical calculations explained. The success of the experiment should be evaluated by referring to the objective of the experiment as presented in the introduction. Was the experiment successful? Were the objectives met? What is the overall significance of the experiment?

The **Literature Cited** section lists all the references used in preparing the report. This section is most formalized of all in its format. Each scientific journal has a slightly different style, which contributors must follow to the letter. Student reports may also be required to follow a certain form. The best way to write this section is with the help of an example. Often college courses use scientific journals as models. The *Journal of Chemical Education*, *Analytical Chemistry* and the *Journal of the American Chemical Society* are examples of chemical journals, which have been used in this way; the *Journal of Organic Chemistry* is often used in organic chemistry courses. When giving references it is important to notice carefully all words that are set in italics or boldface in the example references. Typesetters use different fonts for italics and boldface that are difficult to reproduce when typing or handwriting, though many word processing programs are able to reproduce them. Words that are set in **italics can be indicated by an underline. Boldface can be represented by a wavy underline.** Typically, a reference to a book appears as follows:

REID, R. C.; SHERWOOD, T. K.; PRAUSNITZ, J. M. *Properties of Gases and Liquids*; McGraw-Hill: New York, 1977.

A reference to a scientific journal follows this general form:

LEE, L. G.; WHITESIDES, G. M. *J. Am. Chem. Soc.* **1985**, *107*, 6999.

17.3 Technical Writing

Scientific writing is not limited to scientific journal articles. Scientists on every level are more likely to achieve success if they are able to describe their work and explain its significance to others. Technical writing can vary from a brief explanation of how to use a piece of equipment to a lengthy report on the activities in the laboratory. Technical writers produce articles written for the layman explaining technical subjects in understandable terms. Effective technical writing is a job skill that is very much in demand. College-level assignments that involve report-writing on technical subjects require the same considerations as professional writing.

First, consider the audience. Will the material be read by a skilled professional or a layman? If it cannot be assumed that the reader is familiar with the basic principles of the field being discussed, then the writing must include some basic background information, with special attention given to explaining technical vocabulary that may not be understood by the reader.

Most writing projects begin with a visit to the library to find appropriate source materials. Again, the level of the project will determine how the literature search is conducted. The original research reports contained in scientific journals can be found through indexes such as those provided in *Chemical Abstracts*; using the abstract indexes is a skill that must be developed through practice. Chemistry students usually are given a special course in the chemical literature that includes training in the use of *Chemical Abstracts*. Many science reports, however, require only limited use of original research papers. Science encyclopedias and dictionaries, along with periodicals written for the layman, can provide the background information for a science report and may indicate as well the authors and topics that might be explored in a more detailed technical search. Science and technology encyclopedias useful as sources of background information include:

Harper Encyclopedia of Science
McGraw-Hill Encyclopedia of Science and Technology
Van Nostrand's Scientific Encyclopedia

More specialized encyclopedias include:

Encyclopedia of Chemical Technology
Encyclopedia of Physics
McGraw-Hill Encyclopedia of Energy

Dictionaries can be useful in defining technical terms and concepts. Those that are useful in chemistry-related topics include:

Chamber's Dictionary of Science and Technology (McGraw-Hill)
Chemist's Dictionary (Van Nostrand)
Hanckh's Chemical Dictionary (McGraw-Hill)
McGraw-Hill Dictionary of Scientific and Technical Terms

Facts and data can be found through the many scientific handbooks. Some of the handbooks used in researching chemistry papers include:

CRC Handbook of Biochemistry
CRC Handbook of Chemistry and Physics
CRC Handbook of Environmental Control
Merck Index

Review articles in periodicals like *Scientific American* give useful information on a variety of scientific topics. They can be conveniently found through the *General Science Index*, which provides a comprehensive subject index to English language periodical literature in the sciences. On-line computer search services are increasingly used to locate periodical references. A major resource of the library not to be neglected is the expertise of a good science librarian.

Technical writing depends no less than any other form of writing on the basic language skills of the writer. Incorrect spelling and grammar can mar the effect of the most interesting and original narrative. A good guide to English usage belongs next to a dictionary on the writer's desk. Good writing style is developed through practice in writing and rewriting. A clear, direct style contains no unnecessary words. Consider the following example:

```
At this point in the experiment the mixture was
heated up through the use of a hotplate.
```

A much improved version is:

```
The mixture was heated with a hotplate.
```

Some science publications prefer that use of the first person ("I heated the mixture") be avoided. Use of the passive voice "the mixture was heated" is then indicated. In other uses the more direct form of the active voice may be preferred, as in, "We decided to heat the mixture" rather than, "It was decided that the mixture should be heated." When writing instructions the imperative is often a good choice: "Heat the mixture on a hot plate" is more direct than "The mixture should be heated on a hot plate."

There are many references available to you to help you develop the valuable skill of communicating information. General references include:

W. STRUNK, Jr.; E. B. WHITE. *The Elements of Style*, Macmillan: New York, 1979.
MARGARET SHERTZER. *The Elements of Grammar*, Macmillan: New York, 1986.

References pertaining to technical information are:

B. EDWARD CAIN. *The Basics of Technical Communicating*; American Chemical Society: Washington, DC, 1988.
ANNE EISENBERG. *Writing Well for the Technical Professions*; Harper and Row: New York, 1989.

17.4 A Notebook Example

The following three pages illustrate the proper manner in which a laboratory notebook should be kept. These pages are reproductions of a student's analytical chemistry laboratory notebook. The style of this notebook conforms to the guidelines presented in Section 17.1 of this booklet.

Before entering the laboratory the student had written the introduction and experimental section. A section of data/results was begun as the student collected information. (Note that the student recorded the weights of several tablets as well as the time required to complete the coulometric titration of the ascorbic acid tablets.) This entry is reproduced on page 107.

Next, the student was required to perform several calculations with these data. They were begun in the laboratory and are found on the next page (p. 108). (Note that this page would be the reverse side of the data entry page. It would be the left hand page of the notebook.) Normally, the right hand side of the notebook is used for data entry and analysis, while the left hand side is used for "scratch" calculations. This method eliminates the need for "loose" paper for initial calculations and other scribbles.

The last page (p. 109) contains the conclusions. Note that the student corrected the calculated values and his grammar by drawing a line through the unwanted entry. Erasure or correction fluid was not used.

This simple system of using the different sides of the notebook for formal entries and simple calculations can greatly enhance your ability to organize your data and to present your findings.

Coulometric Titration of Ascorbic Acid with Iodine 3/23/00

Introduction: The purpose of this experiment is to determine the weight percent of ascorbic acid in a vitamin tablet.

Experimental: Add 0.1M KI solution and a spatula of soluble starch. Cautiously add 1sec increments to the current until a pale blue color appears. This will be the matching color. Crush a vitamin tablet and weigh 20 to 30mg to the nearest 0.1mg; transfer to the cell. Turn on current. Repeat the procedure with 2 more samples from the tablet. Continue to analyze other tablets.

Data/Results: multiplier ~~0.5~~ 0.05 = 48.25mA

250mg tablets / 361mg tablet #1, 364.8mg tablet #2

sample	1	0.0232g tablet	389.2 sec
	2	0.0225g	368.4 sec
	3	0.0243g	——
	4	0.0249g	385.0
	5	0.0240g	359.1
	6	0.0234g	353.0
	7	0.0240g	366.3

Actual Weight	Time	Ascorbic Acid		Weight %
0.0232g	389.2sec	0.0171g	Tablet 1	73.7
0.0225g	368.4sec	0.0162g		72.0
0.0249g	385.0sec	0.0170g	Tablet 2	68.3
0.0240g	359.1sec	0.0158g		65.8
0.0234g	353.0sec	0.0155g		66.2
0.0240g	366.3sec	0.0161g		67.1

Tablet #1 contained on average 263.0 ± 4.3mg ascorbic acid.

Tablet #2 contained on average 244., ± 4.2mg ascorbic acid.

263.0 ± 4.3

244.1 ± 4.2

$$95\%\ CL = \bar{x} \pm \frac{ts}{\sqrt{N}}$$

$$= 263.0 \pm \frac{12.7(4.3)}{\sqrt{2}} \qquad = 244.1 \pm \frac{3.18(4.2)}{\sqrt{4}}$$

$$= 263 \pm 38 \qquad = 244.1 \pm 6.7$$

$$\bar{x} - \mu = 263 - 250 = +13 \qquad \frac{ts}{\sqrt{N}} = \frac{(12.7)(4.3)}{\sqrt{2}} = \pm 38$$

$$\bar{x} - \mu = 244 - 250 = 6 \qquad \frac{ts}{\sqrt{N}} = 6.7$$

<u>Conclusion</u>: The value given on the box for each ascorbic acid tablet was 250mg. ~~We~~ The obtained results for two tablets were $263._0 \pm 4._3$ mg and $244._1 \pm 4._2$ mg. The results are close to the reported value. Some error can be expected in this experiment because the calculations depend on an observed endpoint, and although it is tried, it is sometimes hard to pick the same endpoint each time based on color.

The third run on the first tablet was discarded because the electrode was not in the solution and therefore no reaction was taking place.

The 95% CI produces results of 263 ± 38 for the first tablet, which rounds to $26_3 \pm 3_8$. This is mainly high because it is based on two readings. The second tablet gave a result of $244._1 \pm \overset{6.7}{9._7}$ mg. This is based on four readings.

A t-test on the first tablet gives $+38 < +13$, which says that our result is good and would, on average, be correct 95 of 100 times. Our second tablet gives $-6.7 < 6$, which suggests that our second value can also be accepted.

The overall reliability of Coulometric titrations is good as long as you pick an endpoint and can reproduce it.

Section 18 Chemistry and Career Planning

Chemists often find themselves in careers that seem unrelated to traditional chemistry, yet draw heavily on chemical knowledge and skills. Some chemists go into management; management positions in science-related industries increased by 30% between the years 1994 and 2000. Technical sales, patent law, and forensic science are only a few of the careers open to holders of chemistry degrees. Laboratory workers are needed in materials science, polymers, and biotechnology; all these fields depend on the molecular science of chemistry. The traditional skills of chemical analysis and synthesis are in demand for such areas as environmental testing and pharmaceutical development. A wide variety of rewarding career options are available for those with chemical training. The overall unemployment rate for chemists is very low, averaging about 1%.

18.1 Materials Science

One of the major predicted growth areas in the economy is in the area of materials science. The new superconducting materials that promise major breakthrough applications in fields as diverse as communications and transportation are the product of materials science. The development of new graphite materials is resulting in new types of tennis rackets, aircraft, and auto bodies. New ceramic materials are being researched for a variety of uses, including automotive pollution control. Fiber optic materials are revolutionizing communications.

Like many of the fields that offer exciting new job opportunities, materials science is interdisciplinary, requiring a knowledge of chemistry and physics. The field employs both chemists and chemical engineers. Courses in polymer science, metallurgy, and computer science are also helpful. Industry employs the largest number of materials scientists: government and university positions are possible in this field as well.

18.2 Polymer Science

The long-chain giant molecules called polymers form a special class of materials that find a wide variety of uses. Synthetic fibers like nylons and polyesters are the basis of a major industry. Packaging materials as diverse as Saran Wrap™ and Styrofoam are made of polymers. Televisions, computers, toys, tapes, and CDs all make use of polymeric materials. Increasing concern about solid waste disposal is prompting research on plastic recycling and degradability. A promising research area involves modification of polymer properties to make materials that will be compatible with human tissues for medical transplants. Preparation for an industrial career in polymer science may involve a degree in chemistry, chemical engineering, or polymer science and engineering.

18.3 Environmental Science and Technology

Increasing concern about the environment has created numerous job openings for scientists who want to find solutions to the problems caused by pollution. Understanding the chemical and biochemical reactions that produce and consume chemicals like carbon dioxide and methane in the atmosphere is critical to understanding the possible long-range warming process known as the greenhouse effect. Chemists discovered that the chemical reactions of CFC's, or compounds containing carbon, fluorine, and chlorine, with ozone in the upper atmosphere, were causing ozone depletion and a resultant increase in the amount of harmful ultraviolet radiation that is reaching Earth's surface. Now chemists must discover substitutes for these compounds that will replace them in such uses as refrigerants for refrigerators and air conditioners. Automobiles and factories must be designed so that their combustion processes do not foul the air. Factory effluent must be treated or recycled so that streams and groundwater are not polluted.

State and federal agencies such as the Environmental Protection Agency employ chemists to monitor pollution and help to find ways to decrease its sources. Waste management companies recycle materials or dispose of them responsibly, an increasingly technical task. Chemists, biochemists, and chemical engineers all find employment in areas of environmental technology.

18.4 Biochemistry and Biotechnology

Biochemists study the chemistry of living systems. Understanding the reactions, which occur in the human body, in animals, plants, insects, viruses, and microorganisms makes new approaches to curing disease and improving food technology possible. New information from the human genome project will have profound implications for our society. Colleges and universities employ almost half the biochemistry workforce; the rest are employed by government agencies or private companies.

Biotechnology is a burgeoning new interdisciplinary field employing biochemists, chemists, and biologists. Most of these are employed in industry, either by established pharmaceutical and agricultural companies or by new biotechnology venture firms in this rapidly growing area. Through biotechnology, scientific breakthroughs in molecular biology are used to develop new products for the commercial marketplace. Microorganisms are being produced on a large scale, which are able to produce insulin or the human growth hormone. Specific microorganisms are being produced with the goal of consuming oil spills or hazardous chemical waste. Genetic variation of plants may produce varieties, which are richer in nutrients or resistant to insects. These are but the beginnings of the commercial applications, which may be expected in biotechnology.

18.5 Medicinal Chemistry and Clinical Chemistry

Biochemistry and biotechnology are improving both our understanding of diseases and the methods of treatment, but they are not the only fields of chemistry related to medicine. Medicinal chemistry and clinical chemistry are specialties in which chemists have a direct impact on health care.

The medicinal chemist develops new therapeutic agents. Chemical compounds are designed and synthesized with the aim of producing molecules, which will act on one area of the body, such as a certain part of the brain. Once the compound has been synthesized, it is ready for the long series of tests for both positive and harmful effects, which all new drugs must undergo. Tests may indicate that further modification of the molecule's structure is required. To prepare for this work, which links chemistry, biology, and medicine, requires training in chemistry, pharmacology, or medicinal chemistry.

The clinical chemist applies the techniques of analytical chemistry to body fluids. A variety of chemical instrumentation is used, and the modern clinical laboratory makes extensive use of computer automated testing. Administrative duties may form an important part of the clinical chemist's work. Clinical tests are performed both to monitor normal body functions and to test for therapeutic and toxic drug levels in the body. Often critical decisions about health care are based on laboratory results. Many clinical laboratories are involved in research-related testing, to find the results of new procedures, drugs, and equipment or to support basic research. Clinical chemists are employed by hospitals, universities, government and industry.

18.6 Forensic Science

Forensic chemists apply the skills of analytical chemistry in the crime laboratory. Suspected samples of drugs must by analyzed for authenticity by chemical instrumentation if they are used as legal evidence. Body fluids and samples of body tissue are provided by the medical examiner for analysis in the case of a homicide, and cases of sexual assault involve laboratory testing as well. Traces of blood or of gunshot residue can be revealed by chemical tests. Other skills are required of the general forensic scientist, such as investigation of the crime scene. Though these are not directly related to chemistry, some of the most eminent forensic scientists were trained as chemists. Most forensic scientists are employed by government laboratories.

18.7 Radiochemistry

An exciting area of chemistry, which is experiencing a severe shortage of trained workers is radiochemistry. This specialty is needed in a multitude of application areas, some of which are growing rapidly.

Nuclear medicine is seeing rapid advances in such areas as the use of monoclonal antibodies in cancer research. The radiopharmaceutical industry needs radiochemists both to manufacture radionuclides and to oversee waste disposal and protection from radiation hazard. Positron emission tomography (PET) is a diagnostic application of radiopharmacology, which requires radiochemists to staff PET centers and to develop and provide radionuclides and labeled compounds. PET is expected to expand rapidly as its use shifts from primarily research use to more routine medical applications.

Nuclear power production requires radiochemists both in power plants and in related industries to oversee the running of plants and to supervise waste treatment.

Environmental chemistry makes use of radiochemistry in several ways. Accidental emissions from power plants such as the Chernobyl accident require skilled monitoring to assess environmental and health effects. Radionuclides can be used as tags to monitor complex systems such as the buildup of greenhouse gases or the sources of atmospheric CFC's, which destroy the ozone layer. Neutron activation analysis can analyze for trace elements in rocks and help explain Earth's history.

18.8 Patent Attorneys and Patent Agents

A background in chemistry can serve as preparation for the career of patent attorney or patent agent. A patent attorney must have a law degree as well as a science degree. A patent agent may have a degree in chemistry, physics, engineering, or related technological fields. For certification, the patent agent must pass an examination on patent procedures and rules. The patent agent examines the patent literature to determine whether a client's invention is patentable and writes patent applications. Patent lawyers may in addition represent a client in litigation on licensing, trademark or copyright issues, trade secrets cases, and antitrust cases. Patent agents and attorneys are always well-paid and in short supply. Their work involves communication skills as well as scientific training. They are in demand both by government offices and private law firms.

18.9 Chemical Education

Chemistry is taught at both the high school and the college level. College chemistry is taught at two-year and four-year colleges and at universities. Teaching positions at larger universities are likely to involve greater emphasis on graduate education and research than on undergraduate instruction. Demand for high-school teachers is increasing in most areas of the country.

Teachers at all levels speak of the satisfaction involved in interacting with young minds and influencing future careers. Individual freedom is often greater than in many other types of employment, and scheduling is more flexible. College professors have many obligations in addition to teaching, however. Committee responsibilities are required as a part of participation in college governance, and administrative duties may be part of some professors' workloads. Research involves not only study and planning, but also the training of graduate students and postdoctoral fellows and the writing of proposals for research funding. Financial rewards may not be as great as those of the industrial chemist, especially at the high school level and at smaller higher education institutions. For those who teach, the quality of life in the work experience is often a major part of the reward.

18.10 Chemical Information Careers

Chemical information is produced at an increasingly rapid rate as new discoveries are made and the results published in scientific journals. Several career possibilities are available to the chemist who can help others to locate and understand the relevant chemical information for their needs.

The chemical information specialist is familiar with both the chemical literature and the on-line computer services through which chemical information is increasingly accessed. Libraries use chemical information specialists, but chemical companies need them also to help their researchers. Often toxicological and environmental information is needed as well as experimental data retrieval. This special kind of problem-solving expertise should remain in demand as information retrieval becomes an increasingly complex field.

Abstracting services summarize and index chemical discoveries so they can be found by their researchers. They employ chemists to prepare information both for written abstract material and for on-line search services.

Science writers are needed on all levels, both to edit and write scientific publications for the scientists and to interpret scientific information in terms the general public can understand. Research organizations, medical centers, technical companies, and government agencies all need writers who can communicate clearly with a variety of audiences. In addition to print media, radio and television employ experts in science communication.

18.11 Other Chemistry Careers

This list of some current options in chemistry careers is by no means complete. Chemists work in museums and metallurgy plants, in food science laboratories and consumer testing services, in adhesives research and photography laboratories. The materials of our modern world are developed by chemists, and chemists monitor the safety of these products and of our environment. The materials of our future and discoveries, which will prolong our lives and improve the quality of life will be the products of chemical research.

Section

19 Study Aids for Chemistry

Chemistry is a field of science whose basic foundations and concepts go back for many centuries, if not a millennia. All of this activity by scientists has produced a great body of knowledge, which is collectively called "chemistry." The vastness of this body of knowledge is more than one individual could ever know. Even with a lifetime of study, no one person could ever be well learned in all the areas of chemistry.

However, the study that you are to embark on this year with the text, *GENERAL CHEMISTRY*, is a well-organized guide into the basic concepts of chemistry and the connectivity that chemistry has to every facet of modern life. Over the coming year, you will be exposed to discoveries and wonders in chemistry that now define our modern life and offer ways to improve the quality of life for all people.

Having said this, let us emphasize that the material presented in this course will not require all of your waking hours to successfully comprehend and master. You have probably heard "horror stories" from older students concerning chemistry. As a chemistry teacher (and one who has taught thousands of students), let me say that their stories are quite overrated. It has been my perception that the students who do not do well in chemistry are often the ones who have not learned how to study effectively.

The habit of good studying is not one that comes naturally. It is a learned process. It is not a hard process; in fact most of the basis of these habits are founded in good "common sense."

On the next few pages, you will find suggestions on how to develop good study habits. Also listed are suggested ways to get the most out of your textbook, ways to relax while studying chemistry, and tips on taking exams. You may already be practicing many of these techniques. If you see new ones, try to incorporate them into your study routine.

Life is a learning process. While the focus of your learning will change as you go from school to a job, the habit of good studying really does not change. Good study habits can make you a "quick-learner."

Most of all, good study habits can help you enjoy chemistry. That's right, enjoy chemistry! It is truly filled with wonder and amazement. It is rich in history and is relevant to your everyday life. The understanding of the concepts presented in your text and reinforced in this booklet will most certainly allow you to see the connectivity of chemistry to all that is around you. While good study habits will allow you to go beyond the memorization of the material and into critical thinking, it is the skill of critical thinking that sets the stage for your enjoyment of chemistry. This enjoyment from your understanding of chemistry can be truly wondrous.

19.1 Good Study Habits

Good study habits are not among our "basic instincts." Most of us have to work long and hard to develop good study habits. Here are some concrete things you can do to promote your success as a student.

- Set a long-term goal, that is, a career goal. Goals keep us going when the going gets rough.
- Set short-term goals, that is, daily or weekly goals. Short-term goals motivate us and keep us on target.
- Make a realistic daily, weekly, and term-long schedule. Post copies of your daily schedule where others can see it. That will force you to stick to it. Also schedule relaxation time for yourself! We all need that!
- Observe "good students," and make yourself look and act like a good student. Soon you will start thinking of yourself as one. We often have to "act" ourselves into a different way of thinking.
- Create a study area that meets your needs. The study area should have few distractions and be quiet. Try to keep it uncluttered. Having "a place" to study helps you get in the mood to study.

- Select the best time for you to study. We all have different "internal clocks." Study when you are most alert.
- Reflect and review daily. *Reading a lesson and doing problems is not studying*. You read the lesson and do the problems to gain the knowledge you need to study the material. Studying involves moving material from memory to understanding.
- If you can't explain a concept in your own words and relate it to something you already know, you probably don't understand it and won't remember it very long.
- Get to know others in your class, and form a serious study group. Then study together at least once a week. It will keep you "honest." It is easy to slip into sloppy study habits and fool yourself into thinking you know more than you do. Having to explain ideas in a study group will bring you back to reality very quickly. We learn and learn how to learn from working with others.
- Work to stay relaxed while you are studying.

19.2 Reading Your Textbook

If your assignment is to read pages 88 to 108, the worst thing you can possibly do, aside from not doing the reading at all, is to open the text to page 88 and begin reading. Reading an assignment is a three-phase process. It has a *before*, *during*, and *after* phase, and each is equally important!

The Before Phase

- Skim the entire assignment first so you have a general idea of the content, and try to relate the content to concepts you already know.
- Look at your syllabus and see how many lectures are devoted to the assignment. See what other material is included with this assignment on the next test.
- Judging from the length and the difficulty of the reading assignment, try to determine how long the task will take and plan ac-

cordingly. (Remember that reading technical information often takes more time than reading nontechnical works.)

The During Phase

- Mark up your text. Write notes in the margins. Highlight key concepts. (Try not to buy a book that has been marked up too much. You will be distracted by someone else's markings.)
- If examples do not show all the steps, write in the missing steps so you will remember how to do the problem when it is time to review for a quiz or test.
- Look at the pictures, figures, and graphs in the text. Remember, Alice in Wonderland asked, "What good is a book without pictures?" Fortunately, most modern chemistry books are filled with pictures. Take advantage of them. Remember that chemistry is visual. When the text refers to a drawing, figure, diagram, chart, photograph, etc., look at it right away. Study it. Sometimes one picture can be worth many words!
- Spend time reflecting on what you read. Can you put the concepts into your own words?
- Make a short outline of the material you have just read. Placing new concepts into your own words helps you to retain and master the new material.
- Do you need to hear yourself reading the material? Reading difficult material aloud slows you down and lets you hear as well as see. Two senses are better than one! Remember that you are reading for understanding.

The After Phase

- Go back to any sections you did not understand. If after going over them again, you still have concerns, put a question mark beside them and ask about them during the next class or help session or meeting with your study group.
- Go back over the entire reading assignment, looking at main headings and the notes you made in the margins. Does this information

remind you of the main ideas and content of each section? If not, add to it.

- Write a brief summary of what you read. Include any type of diagram that will help you remember the material. This summary, what you have underlined in the text, what you have written in the margins of the book, and the notes you have taken in class are the materials you need to use when you review daily and study for quizzes and tests.

19.3 Learning to Relax

It's hard to study effectively, and you certainly won't perform your best on an exam if your body is overly tense. Also, when you are tense, things always look worse than they are. However, some stress is normal when you enter into the unknown. And a little tension keeps you from getting too complacent and making careless errors. It is the excess tension that you need to be able to control.

There are three factors you can consciously control to reduce your stress and increase the effectiveness of your studying: posture, muscle tension and breathing. We will discuss just a few techniques you can try with each factor.

Posture

When you are concentrating hard, especially when you are tired, you often tend to pull your neck down into your shoulder blades, collapse your chest and let the lower spine bow out. Sitting with the body in this position tends to make you tired and tense. Muscles work overtime to hold the body up, and the lungs cannot breathe efficiently.

It is important to have good posture while sitting and studying. If you cannot hold a good posture, you may need to take a break and relieve muscle tension.

Muscle Tension

A tense muscle uses more energy and requires more oxygen and nutrients than a muscle at rest. But a tense muscle constricts the capillaries that deliver oxygen and nutrients to it, which leads to fatigue of the muscle. The exhausted muscle becomes even more tense, starting a downward cycle. There are three things you can do to relieve muscle tension: stretching, massage, and movement.

- While you are studying, remember to stretch. Stand up periodically and stretch your body.
- Take a walk. Walking is one of the best ways to relax. Even a short walk around the block will help you relax and feel more alert.
- Periodically look out the window or across the room to relax your eyes.

Breathing

Your breathing is tied to your emotions. Have you noticed that after you have been scared, your breathing becomes rapid and shallow? This is one symptom of what is known as "fight or flight," a remnant of the days when we had to either confront a danger or run. In the modern world, this reaction does little to help us.

The same "fight or flight" reaction can happen on a smaller scale when you are studying unfamiliar material or when you are taking an exam. Here are some breathing techniques you can try anytime you feel tense.

- Breathe in slowly through your nose as you count to five.
- Then breathe out slowly.
- Pause for a second when you have completely exhaled, then repeat for a few more breaths.

When you inhale, expand your abdomen first to fill the lower lungs. Then let your chest rise to fill the upper lungs. Don't force the breath; let it come easy and natural. You can do this exercise a few minutes before you take an exam.

19.4 Conquering Tests

Preparing for the Test

- Test preparation begins on the first day of the course. Keeping current with all daily assignments is the most effective method of preparing for tests.
- Form a study group. You never really learn anything until you have to explain it to someone else. A study group will force you to do this. Working with others will also help you learn more study strategies. But study on your own both before and after the study group.
- Tests are more difficult than individual assignments because they contain a greater variety of material. To avoid getting confused, you must "overlearn" the material. To "overlearn," study actively until you really feel that you understand the material or can work the problems with no difficulty. Check to see how long you had to study or how many problems you had to work to reach this point. To "overlearn," you must spend at least half as much more time or work half as many more problems. Then you must also spend a small amount of time each day, maybe 10 to 15 minutes, reviewing past assignments after you have completed the new assignment.
- If there is a concept you do not understand, get help immediately. In chemistry, learning is often sequential; i.e., new lessons build on past lessons. Therefore falling behind can be fatal!
- Try to predict test questions as you are studying.
- If practice tests are available, get them as soon as possible and do them at least several days before the test. Then you will still have time to get help if you find there are questions that are giving you difficulty.
- If you do fall behind in your studies, never try to learn new material the day or the night before a test. You will not have time to

"overlearn" it. Therefore, what will normally happen is you will confuse new material with what you already know. This confusion can cause you to miss even more test questions than you would have missed had you not tried to "cram in" the new material. Concentrate on what you know on the night before the test. Then, after the test is over, catch up immediately.

- The night before the test, be sure to get enough rest. Also check your calculator and put it and any other materials you will need in a place where you will be sure to pick them up before leaving for class.

Taking the Test

- *Wear a watch!* Bring your calculator, and make sure it is functioning.
- Get to class a few minutes early so you have time to get comfortable and get organized before the test is handed out.
- Put your name (and your student ID number) on the test!
- As soon as you get the test, write down on the test paper any information you are worried about forgetting. Then you won't have to worry anymore, and your mind will be free to concentrate completely on each test question.
- Preview the entire test. It only takes a minute or two but can tell you much. Previewing tells you, the number and types of questions the test contains (multiple choice, true-false, matching, essay, problems, etc.), the number of points each question is worth, any additional information you need to write down to be sure you remember it, and how to schedule your time.
- Previewing should also build your confidence because if you are prepared, you will know immediately that you can answer most of the questions.
- Read any directions carefully and completely. If there is something you do not understand, raise your hand and ask about it. The instructor will probably be able to clarify it for you. In any case, the worst that can happen is the instructor will say he or she cannot answer your question.

- Begin with the questions that are easiest for you. This strategy will build your confidence and lessen your anxiety. It may also remind you of things you need to know to answer the more difficult items. In addition, it will prevent you from losing points by spending too much time on things you don't know and running out of time before completing items you do know.
- Return to questions you skipped. If they are objective questions, i.e., true-false, multiple choice, matching, etc., make an "educated guess." (See "Mastering the Multiple-Choice Test.") If there are essays or problems to be solved, do as much as you can. If the problem has several dependent parts and you do not know how to do the first part but do know how to do the next part, write a note saying, "I'm not sure how to do part A, so I am going to assume the answer is (make up a reasonable value) and use this value to complete the problem." Remember that part credit is better than no credit! Answer every question as completely as possible.
- Be neat, and show your work; it will be easier to check your answers.
- Check your work! If you only have a few minutes left, check the first few questions you answered. You were most nervous then, and therefore you might have made careless errors. Then check the problems you thought were the easiest. You also make careless errors when you are too relaxed. If you eliminate careless errors, you can usually increase your grade by at least 10%!
- Do not get nervous because others are leaving before you. Those who finish early often are the weakest students. Use all the time allowed. There is no prize for finishing early!

Mastering the Multiple-Choice Test

In addition to the general test-taking strategies presented in this booklet, there are some specific things you can do to increase your score on multiple-choice tests.

- As soon as you get the test paper, jot down any information you know you will need and are afraid you will forget.

- Scan the entire test. If any of the questions remind you of any other facts you are afraid of forgetting, write them down.
- Read the directions. Most multiple-choice tests have only one correct choice, but don't assume anything.
- Answer the easiest, least time-consuming questions first. This strategy will assure that you have time to answer all the questions you know. It will also build your confidence and remind you of things that might help you answer the more difficult items.
- Before you look at the answer choices, work the problem and ask yourself, "Does this answer make sense?" Remember that many of the incorrect answer choices will be derived from mistakes that students commonly make. If you see your answer as one of the choices, don't immediately mark it.
- If you don't find your answer among the choices, look again. Perhaps your answer is there but in a different form. For example 1/2 could be written as 0.5, 0.50, or 50%, etc. Maybe the answer is in scientific notation, and you calculated your answer in decimal form.
- After you've answered all the questions you know, carefully reread each question you skipped. You may find you now remember how to do them. Doing the questions you know warms you up, and sometimes answering one question will spark your memory on how to answer another.
- If you are still unable to answer some of the questions, make an "educated" guess. Can you eliminate one or more of the choices? For example, if the question asks for grams, eliminate all the choices that don't have units of grams.
- Never leave a multiple-choice question blank unless the directions say there is a penalty for guessing.
- Use all the time allowed. There is no prize for being the first to finish.

Section
20 Math Skills Test Answers

answer		if you missed this problem, refer to:
1.	3	Section 2.1
2.	13.019	Section 2.1
3.	0.103	Section 2.1
4.	3.56×10^{-4}	Section 2.2
5.	25.19	Section 2.3
6.	1.1	Section 1.2
7.	66.1°	Section 1.2
8.	0.0155	Section 1.1, 1.2
9.	8.06%	Section 2.2
10.	1.6	Section 1.1, 4.1
11	0.68	Section 1.1, 4.4
12	439	Section 1.4
13.	2.03	Section 1.4
14.	1.6	Section 6.2
15.	0.157	Section 4.4
16.	-1.77	Sections 1.3, 9.2
17.	12	Sections 1.3, 9.2
18.	34	Sections 1.3, 9.2
19.	31	Section 9.2
20.	984	Section 9.3